D1163567

FIRESIDE

The Complete Book of WARGAMES

BY THE EDITORS OF
Consumer Guide®
with Jon Freeman

A Fireside Book
Published by Simon and Schuster
New York

A Fireside Book
Published by Simon and Schuster
A Division of Gulf & Western Corporation
Simon & Schuster Building
Rockefeller Center
1230 Avenue of the Americas
New York, New York 10020

Designed by Irving Perkins
Manufactured in the United States of America
1 2 3 4 5 6 7 8 9 10
1 2 3 4 5 6 7 8 9 10 Pbk.

Library of Congress Cataloging in Publication Data
Main entry under title:
The Complete book of wargames.
(A Fireside book)
Includes index.
1. War games. I. Freeman, Jon. II. Consumer guide.
U310.C65 793'.9 80-72

ISBN 0-671-25374-3
ISBN 0-671-25375-1 Pbk.

For contributions of game evaluations and some introductory material, the editors would like to thank Richard Berg; Jeff Johnson, who also designed the introductory game *Kassala;* Dave Minch; Jim Connelley, John Prados; and Tony Sabado.

Contents

PART 1

An Introduction to Wargames

CHAPTER 1

Can War Be Fun?

PHYSICAL CONFLICT has been with us at least since Ug the caveman hit Mug over the head and dragged off Raquel Welch by the hair. (Some historians claim it was Carole Landis. No matter.) War, strictly speaking, had to wait more formal tribal organization—a government and the beginnings of "civilization."

Wargames probably originated soon thereafter. The remains of board games have been found in the ruins of ancient Ur, and there are several versions available today of a game called Senat or Senet, which was played in the days of the boy-king, Tutankhamen. Pachisi (*Parcheesi*) may be as old as "Tables," a Roman ancestor of backgammon. No one knows how old chess is, but it is thought to have evolved over many centuries from something like a modern simulation: before the process of abstraction went so far that the game became only a pale shadow of a real battle, chess pieces probably represented different sorts of troops whose movement and combat capabilities were reflected, more or less realistically, on the board. Chess variations designed to add realism were popular throughout the Middle Ages.

The practical beginnings of modern wargaming are linked to a man named Helwig, Master of the Pages for the Duke of Brunswick. Although played on a board with squares (1,666 of them!), his game, devised in 1780, was closer to the modern game *Feudal* than it was to chess, and it had a number of features found in our contemporary wargames: different types of units (infantry, cavalry, and artillery), movement allowances that varied by unit type, colored terrain, and special rules for fortresses, pontoons, and entrenchment.

A few years later, Georg Vinturinus "improved" Helwig's game. He increased the number of squares to 3,600 and added pieces, special rules, and record-keeping. Vinturinus' version was set on a map depicting the Franco-Belgian border, one of history's more popular battlefields (the late-eighteenth-century version of the eastern front).

After Vinturinus, wargaming took a lengthy detour, becoming less an entertainment for the military-minded and more a training technique for the military. A Prussian named Von Reisswitz replaced the game board with a sand table; a hill was no longer a colored square, but a tiny mound of sand or earth on a scale of one foot to about half a mile. Since there were no squares to regulate movement, the pieces were moved according to rulers and gauges. This is essentially the modern miniatures approach.

In 1824 Reisswitz's son, a lieutenant in the Prussian Army, played Vinturinus to his father's Helwig by using realistic military maps drawn to an even smaller scale (one foot equaled about a mile and a half), by following this scaling-down process even more strictly than his father had done; and by incorporating still more special rules and tables. His most significant contribution, however, was the use of limited military intelligence: the players were not allowed to see the total situation or to know their opponent's exact aims and disposition. Managing this, of course, required an umpire—a neutral party who kept each side supplied with only the information that their miniature forces could logically obtain. This was taken to such an extreme that one portion of an army was not permitted to act on the basis of intelligence received by another faction until enough time had lapsed for a "courier" to travel between them.

General Von Muffling, the Prussian Chief of Staff, saw the value of the game as a training device and ordered the army to adopt it. He and a later Chief of Staff, Helmuth von Moltke, continued to support the concept; and over the succeeding decades, despite a variety of changes and adjustments that made the game ever more complex, variations of Reisswitz's game became fixtures in the Prussian Army.

After the Prussians' stunning victory against the French in 1870, other armies throughout the world rushed to adopt Prussian equipment, dress, organization, and training methods—including *Kriegspiel* (the "wargame").

Confederettes—Heritage's cast metal historical figurines for displays, dioramas, and gaming. These exquisitely detailed figures are only 15mm tall! Other lines are 25mm, 54mm, or 75mm tall, and cover all eras of history, as well as science fiction and fantasy (such as J. R. R. Tolkien's Lord of the Rings*). Heritage also publishes booklets describing games with these and its other figures.* (PHOTO: HERITAGE MODELS, INC.)

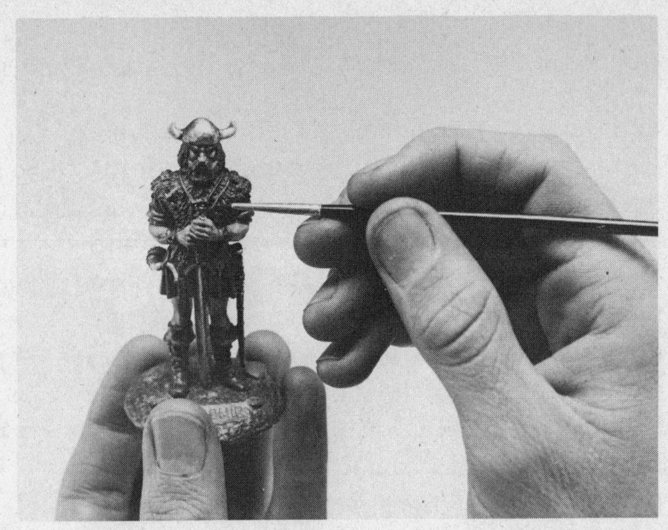

Painting a miniature figure by hand. (PHOTO: HERITAGE MODELS, INC.)

In 1876 Colonel von Verdy du Vernois sounded a note that was to echo all the way to the present. Wargames, he complained, had become so complex and cumbersome that they weren't really playable any more. His solution was to give even more freedom and responsibility to the umpire, who would be an officer sufficiently versed in the art of warfare to run things realistically without having recourse to so many printed tables, charts, and rules, and the endless die-rolling that went with them.

The colonel's innovations caused an international split between wargamers who favored the more traditional "rigid" kriegspiel and those who preferred the new "free" kriegspiel. Over the next thirty years, enthusiasts on both sides of the Atlantic, including Americans William R. Livermore (*The American Kriegspiel*) and Charles A. L. Totten (*Strategos*), contributed to this war between wargames, but in retrospect the differences between the two kriegspiels seem more theoretical than real. Even Reisswitz's original "rigid" game is very "free" indeed by modern standards.

The rift between the two kriegspiels is overshadowed by two far more significant divisions that characterize wargaming in the last seventy-five years. The first is the development of civilian wargames that are almost totally different in design and function from those used by the military. The second is the split in civilian wargaming caused by the relatively recent arrival of the modern commercial board game.

Although strategic in scope, military gaming through World War II was essentially an outgrowth of the Reisswitz kriegspiel. Its purpose was to devise and test actual battle plans that could or would be implemented in the event of war. In this, it was less than completely successful.

To cite an early example, the basic postulates on which the German Schlieffen Plan was based also underlay the wargames used to test it. Since a conclusion is no better than its premises, the results of the games could do nothing except confirm Graf von Schlieffen's already set opinions. In the event, owing to unforeseen or incorrectly forecast events, the anticipated quick victory did not occur, and World War I degenerated into the bloody stalemate of trench warfare.

A world war later, the Japanese tested, with a game, their plan for what would become the Battle of Midway. Using what they viewed as the most likely American forces and tactics, the Japanese suffered severe losses. Instead of changing their plan, however, the officers in charge disregarded this "unacceptable" result and proceeded with the game as if their warships had not been sunk. Successful on their second try, they congratulated themselves on their masterful strategy and put the plan into operation. The actual Battle of Midway in 1942 confirmed in amazing detail the initial results of the game, and the Japanese Navy suffered a major defeat.

These anecdotes illustrate the subordination of wargames to the prejudices of the policy makers. If the results of a game agreed with the preconceptions of those in charge, well and good; if not, the game was simply disregarded. This my-mind's-made-up-don't-confuse-me-with-facts attitude, typical of an authoritarian mentality, is not unique to the military. Despite fears sometimes expressed on these grounds, there is no evidence that wargames, by "supporting" an eager general's policy of aggression, have ever led directly to a war that would not otherwise have begun.

On the other hand, wargames have been helpful at a lower level of decision making. In one of the most famous instances, a game version of the Battle of the Bulge was being played by the staff of the Fifth Panzer Army when the actual conflict erupted. The game was continued, utilizing up-to-date information relayed from the front, and the results were used to direct the German operations.

As the wargaming practiced by the military establishments of the world became more strategic and ever more specifically oriented toward real plans and policies, it reappeared on the tactical level as a game and entertainment for civilians after a hiatus of almost a century. On the eve of World War I, H. G. Wells published *Little Wars*, which contained rules for re-creating warfare on land using miniatures—carefully made and authentically painted toy soldiers—on something resembling Reisswitz's sand table. The book received wide notice and became the foundation for nearly all the miniatures rules in use today.

Between the two world wars, naval historian and fantasy writer Fletcher Pratt published a set of naval rules that did for war at sea what Wells had done for combat on land. Pratt also reinforced a connection between games and science fiction and fantasy that has become even more explicit in recent years.

For forty years, the Wells-Pratt approach—miniatures—was the only form of the hobby in existence. Then, in 1953, Charles S. Roberts published a game called *Tactics*. Like Helwig's game back in 1780, *Tactics* was played on a board with a square grid, color-coded to indicate the type of terrain. It had pieces or counters of heavy cardboard on which were printed combat and movement factors—numbers that indicated how strong the pieces were when they fought other pieces and how far they could move. His mapboard, like Helwig's and unlike that of Vinturinus, had no specific referent; the pieces represented units in two contemporary but hypothetical armies. This first commercial wargame, on which nearly all modern board wargames are based, was distributed by the Stackpole Company, which twenty years later published the first book on commercial board games.

In 1958, Roberts decided to go into the business of producing adult games. The initial offering of his new company, Avalon Hill, was *Gettysburg*, the first modern board wargame on a historical subject. Although *Gettysburg* and *Tactics II,* a revamped version of his original game, used squares on their maps, his *D-Day* did not. By allowing even movement in all directions, a hexagonal grid, an idea that Roberts had "borrowed" from the RAND Corporation, eliminated the problems of squares and moving pieces diagonally.

Hexagons, or "hexes," became a wargaming hallmark, and the Roberts board game approach came to dominate the wargaming hobby in the United States.

There were several reasons for this. The most obvious was that a game of the old Wells variety (with miniatures) required an expenditure of hundreds of dollars for the figures involved, hundreds of hours to paint them properly, and the space for a Reisswitz-type table covered with replica terrain to put them on. After all that, what you had was suited for only one battle—or, at best, for several engagements of one war. Going from Napoleonics—soldiers of the Napoleonic Wars of the early nineteenth century, one of the most popular periods for miniaturists—to, say, the tanks of World War II required a similar expenditure of time and money.

This difficulty led to further specialization within an already specialized field and further restricted the potential sales of any particular "game"—actually, no more than a lengthy set of rules. Since professional publishing efforts were not often economically feasible, collector/gamers usually made up their own house rules to suit the period of their specialization and even the particulars of their collection.

With the minor exception of wargaming

A miniatures enthusiast paints details on elements of a tiny army. (PHOTO: TSR GAMES, INC.)

clubs, which were few and far between, there was no way for an outsider to test the waters. Much like model railroading, the hobby was made up of and limited to a small, stable group of aficionados.

The approach pioneered by *Tactics*, however, suffered no such limitations. For five or ten dollars, a medievalist could afford to try a game based on the Civil War. More important, for the same price any adult could join the fun on much the same footing as a veteran gamer.

Furthermore, the new sort of simulation was not restricted to re-creations of individual engagements on a tactical level. Simply by altering the scale of the map and changing the unit designations, you could simulate anything from a handful of men assaulting a building (*Sniper!*), to an entire war (*1776*), or even to the exploration and conquest of a galaxy (*Outreach*).

For the first time, it was possible to sell a lot of wargames, perhaps 20,000 copies of a single title. A wargaming industry was

Components of Avalon Hill's classic game Gettysburg. (PHOTO: AVALON HILL GAME CO.)

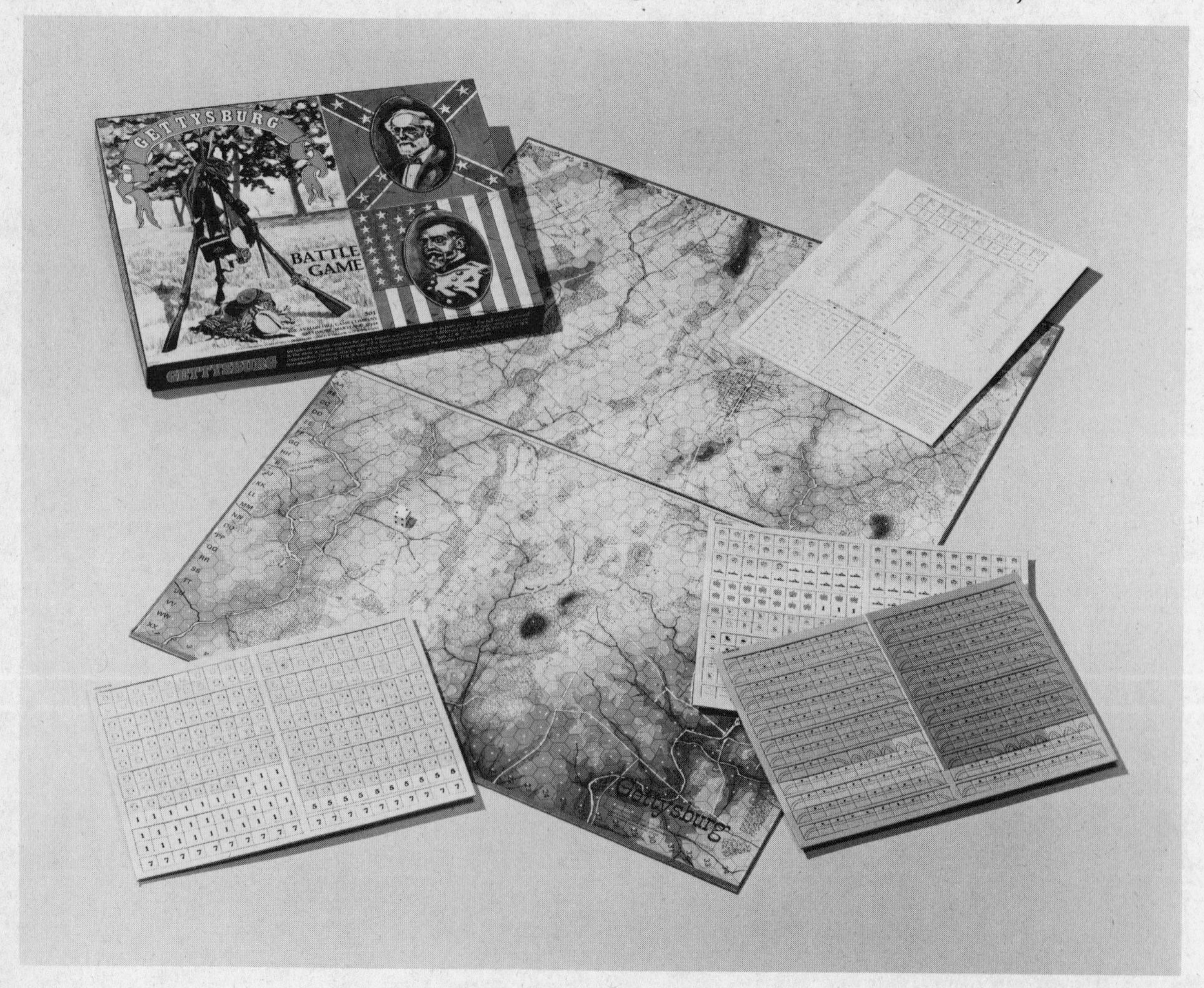

A miniature wargame in progress: ancient armies clash again, but this time with lead soldiers that don't bleed! A hobby as well as a game, miniature wargaming is enjoyed by more and more adults every year. Heritage Models, the largest manufacturer/publisher in the U.S.A., produces extensive lines of metal figures for gaming, booklets of rules or for reference, and complete boxed game sets as well. (PHOTO: HERITAGE MODELS, INC.)

born, and after it came a hobby comprising a large group of people interested in buying and playing wargames. In fact, in a commercial sense, Avalon Hill could be said to have created the entire adult game market.

WAR OF THE GIANTS

Expanding rapidly over the next few years, Avalon Hill published such "classics" as *Waterloo*, *Stalingrad*, and *Afrika Korps*. Then, in 1963, caught in a credit crunch brought about by the shift toward discount stores and the demise of many of its traditional retailers, and facing bankruptcy, the company was taken over by one of its major creditors, Monarch Services, a printing firm. Roberts left, and Thomas N. Shaw became executive vice president and chief operations officer.

The staff at Avalon Hill was pruned, and the company began a policy of using outside designers for many of its new games. It also started *The General* magazine as a house organ to promote its products. The magazine also served as a forum for wargamers to exchange views on the games and their hobby.

From this sprang other lines of communication among hobbyists. Taking a cue from the world of science-fiction fandom, gamers began publishing "fanzines" or " 'zines"—amateur magazines or newsletters characterized by crude reproduction, small circulation, negligible editorial standards, and short lifespans. Home-brewed games of similar quality followed.

Avalon Hill and *The General* magazine had no real rivals until James F. Dunnigan and Redmond A. Simonsen took over one of the better efforts, a magazine called *Strategy & Tactics*, from Christopher R. Wagner and went into business in 1970 as Simulations Publications, Incorporated (SPI). For Avalon Hill, wargames had always been (and, for that matter, still are) only a part of the company's entire adult game line, albeit one to which they devoted the biggest portion of their time and effort. Dunnigan, who had designed *Jutland* and *1914* for Avalon Hill, and Simonsen believed it was possible to make a living producing only wargames. They felt that the potential market was much larger than Avalon Hill had explored, and that it could absorb more—far more—than the two new games a year Avalon Hill was producing.

They were right. SPI simply exploded. Producing more than a dozen games a year, SPI attained something approaching parity with its older rival within three years.

The magazine proved invaluable. *Strategy & Tactics* attracted subscribers with a taste for SPI products, provided free advertising for SPI games, generated steady income, and through a formal feedback system allowed SPI to produce games with a predetermined market. Meanwhile, game sales allowed the company to subsidize the magazine—without which SPI could not have afforded to maintain its high quality or to keep putting a game in every issue. As *Strategy & Tactics* became more of a magazine of military history, game-related material (though not the games themselves, which remained in *Strategy & Tactics* as "historical simulations") was put into a new magazine called *MOVES*, which was more closely akin to *The General*.

From the beginning, Avalon Hill and SPI, a study in contrasts, have been fierce rivals. Initially, SPI was exclusively a mail-order operation, while Avalon Hill distributed its

Cover from SPI's magazine Strategy & Tactics. (PHOTO: SIMULATIONS PUBLICATIONS, INC.)

products almost entirely through regular retail outlets. Although SPI was producing four times as many titles as Avalon Hill, each title sold only one-fourth as many copies. SPI became somewhat notorious for publishing historically accurate simulations that were too dull, unbalanced, complicated, long, or, like *War in the East*, just too big to be playable. Avalon Hill was said to sell games that were easier to understand, more balanced, and more enjoyable, if not entirely historically precise. "Quality versus quantity" was a saying often echoed by SPI's detractors, but few people at Avalon Hill took it seriously.

Many of their differences remain. SPI has larger offices, a larger staff, and a large and active art department. Avalon Hill has fewer offices, a much smaller staff, and not a single full-time artist. SPI uses its own design staff exclusively; Avalon Hill often uses free-lance designers or simply buys a game from one of the smaller companies for redesign and distribution on a much larger scale. Avalon Hill continues to produce nonwar adult games; SPI does not.

Time and the nature of reality have reduced other dissimilarities. In recent years Avalon Hill has encouraged mail-order business, in part by maintaining a small line of games that are available only by mail. On the other hand, SPI games are now found in many stores. In fact, in an effort to sell in still more retail outlets, SPI has begun to use mounted game boards and package games in cardboard boxes, practices it shunned for so long. With the introduction of "folio" games and, recently, "capsule" games, they have also shown a clear interest in playability. In the meantime, Avalon Hill is producing more, and more complicated, games.

Some of the points of contention between the two companies appear to be no more substantial than the controversy over the two kriegspiels. For one thing, many of Avalon Hill's games, from the famous *Blitzkrieg* to the infamous *France, 1940,* were designed by Dunnigan and his staff at SPI. For another, it is clear that SPI has had its successes and the older company its unplayable failures. Today their past habits matter very little. Both companies produce good games, and even their most ardent proponents will buy a game that looks interesting, no matter who published it.

THE THIRD WORLD

While the two giants produce something like three-fourths of the wargames on the market (more, if role-playing games are not included in the tabulation), the competition is getting stiffer all the time.

Gamescience, the first company to enter the lists, preceded SPI. Its most famous game, *Battle of Britain,* sold reasonably well, but the game and the company rapidly disappeared after its founder, Phil Orbanes, who later founded Gamut of Games, sold out to Renwal. Lou Zocchi, the game's designer, then began producing his own games, which dealt predominantly with war in the air. (Avalon Hill's *Luftwaffe* was another of his designs.) Having revived the Gamescience name, Zocchi now distributes the products of other small companies as well as his own.

Simulations Design Corporation (SDC) was explicitly organized as a rival to SPI: it boasted a well produced magazine, *Conflict,* that featured a game in every issue. Its games pointedly emphasized playability, a feature *Conflict* subscribers felt was missing in the SPI efforts of the time. Unfortunately, there were not enough subscribers to sustain the magazine, and when *Conflict* died, SDC went into long-term hibernation.

Game Designers' Workshop (GDW) originally attracted notice because of its lengthy and extremely complex Europa series of interlocking World War II games (*Drang Nach Osten!* and others). With the acquisition in 1975 of John Hill's Conflict Game Co., however, GDW branched out in an attempt to appeal to a wider audience. It now markets quite a diverse selection, from the science-fiction role-playing game *Traveller* to its reasonably simple Series 120 line of games designed to be played in about two hours.

Battleline Publications became known as much for its extra-thick playing pieces as for any other aspect of its solid, well researched games. Some of the company's best designs, like *Submarine* and *Wooden Ships & Iron Men,* were sold to Avalon Hill. In 1978, after becoming a part of Heritage Models, a manufacturer of miniatures, the firm came out with a new line called Gametime Games (*Quest, Sopwith,* etc.), aimed at a more general audience. More recently, to the surprise of most observers, Battleline's two prime movers, Steve Peek and Craig Taylor, left to form their own company, Yaquinto Publications.

Metagaming publishes *The Space Gamer,* a magazine largely devoted to its line of science-fiction and fantasy games. Although some of these games are of standard form and format, Metagaming's fame rests with its microgames: *Ogre* and its successors. MicroGames are about the size of an ordinary paperback book; they typically sell for $2.95, and they are relatively easy to learn and play. In addition to spearheading the science-fiction boom, their success played a significant part in the resurgence of interest in manageable games and, in form and content, led directly to the development of SPI's capsule games.

Unquestionably, the biggest success story in the hobby's last five years is TSR Games. Although this company also publishes a magazine (*The Dragon*) and has repeatedly attempted to diversify, its fortunes continue to depend on permutations of a single game, the extremely popular *Dungeons & Dragons.*

Many of the smaller companies have in common a particular interest in science fiction and fantasy, and a lack of the sophisticated packaging that is characteristic of older and bigger firms. Although the production quality and other aspects of the contents can vary considerably, the standard container here is not the box but the resealable plastic bag or pouch.

Flying Buffalo divides its efforts between solitaire adventures for its role-playing game, *Tunnels & Trolls,* and play-by-mail computer games like *Star Web.* The Chaosium centers its attention on a role-playing game (*Runequest*) and a series of fantasy wargames all set on the imaginary world of Glorantha. Fantasy Games Unlimited emphasizes a variety of rule books for miniatures and role-playing campaigns. Although Automated Simulations plans to expand into all areas of gaming, its first offerings (the *Orion* games and the *DunjonQuest* series) are for microcomputers.

Of late, the wargaming scene is abuzz with change, as the disaffected depart the Establishment, and new companies coalesce around a bright young designer, an interesting game idea, or a new area of specialization.

THE REASONS

"War is hell," said General William T. Sherman, and, while Caesar or Patton might debate the matter, the point is not likely to be argued much today—not with defoliants to destroy plant life, neutron bombs and man-made plagues to eliminate people, and ICBMs with multiple warheads to annihilate *everything*. No, in times like these warfare is regarded as a dirty, nasty, deceitful business.

Who in the world, then, would want to play wargames? And why?

There is no simple answer. A few years ago, the U.S. Army contracted with SPI to design a game for the purpose of training soldiers in small-unit tactics. Contrary to the image of wargamers as militarists, however, the initial proposal generated considerable concern in the pages of *Strategy & Tactics* magazine. Ultimately, though, no one really believed that the game was going to precipitate World War III, or that its absence would be a step toward global peace. SPI accepted the contract and produced the game, which later was made available to the public as *FireFight*. But this is only one game out of hundreds on the market.

The only other connection between the military and modern commercial wargaming is equally insignificant. It would be strange indeed if the armed forces were not interested in wargames. From time to time a group of military enthusiasts, under official sponsorship, does travel to a club or convention to take part in a tournament or in less formal competition. However this is rationalized to the General Accounting Office, participation in such events is at least as much a matter of enjoyment as a test of their grasp of tactics and strategy. Far from being the usual military type, in fact, the "average" player of wargames has never worn a uniform.

WHO WAGES WAR?

Wargamers generally range from fourteen to forty years of age. Gamers generally get hooked in high school or college.

A great percentage of gaming enthusiasts are male. To a small degree, this can be ascribed to discrimination and social hostility. Mostly, though, it's the result of natural selection. Few women seem to have the interest in military history or the open competitiveness of the majority of men who indulge in the hobby. While the growth in the last few years of science-fiction and, particularly, fantasy role-playing games is reducing this inequality, the field remains a male stronghold.

The typical wargamer, if there is such an individual, is a college-educated man from a middle-class background, recently out of school, intelligent, introverted, and technologically sophisticated. He has some interest in military history, plenty of time to devote to a hobby, and a willingness to try complex, challenging games.

This, however, is a gross oversimplification. For every "Joe Average" there are a hundred gamers who depart considerably from the mean. Nor is there a single, simple reason for the appeal of wargames. To get some sense of their attraction, we must examine, individually, the composition and motivations of the varied subgroups which make up the total picture.

The Historian

It has been said often enough that whoever fails to learn the lessons of history is condemned to repeat its mistakes. The historian gamer uses conflict simulations as an extremely concentrated source of information to fill the gaps left by his education. By playing with the variables of tactics and strategy, reinforcement and supply, and timing and preparation, he can gain a unique insight into the crucial factors of an engagement; ideally, he comes to understand why the actual results of a battle or war came about and how they might have been altered.

For this type of person, realism is of paramount importance. If a game is not a reasonably accurate simulation of the actual events, it's worthless, in his opinion. The historian's Holy Grail is a game that will reproduce the results of a real battle—for much the same reasons. Complexity is a concern only in the sense of its appropriateness to the scale and scope of the simulated event and to the degree of the player's interest in the subject. Detail takes time but generally allows him to learn more. Balance of play matters not at all, because the player will usually be using the game in the company of only reference books and maps. He is not looking for an even contest; he's looking for information. A staunch SPI supporter, the historian type of wargamer subscribes to *Strategy & Tactics* but not *MOVES*, buys a game from a lesser company if it's well researched and detailed, but sneers at science fiction and microgames.

The Military Enthusiast

The historian's first cousin is the military enthusiast. Nine times out of ten, this person no more wants to smell the smoke of a real battlefield than the average armchair athlete (whom he closely resembles) wishes to try running through the Steel Curtain of the Pittsburgh Steelers. But as long as no one is getting hurt, as long as things are held a safe distance from reality, the enthusiast finds the tactics and technology of war fascinating. Typically, he loves tanks and prefers tactical games: *PanzerBlitz, Panzer Leader, Mech War '77, Wooden Ships & Iron Men.* Where the historian is concerned only with the past, the enthusiast is as much intrigued by the possibilities of the present and future. Could NATO forces in West Germany survive an attack by the Warsaw Pact countries? How would the F-14A Tomcat do against the MiG Foxbat? What might combat on another planet be like in 150 years? The military enthusiast is concerned with realism only as it affects performance: he'll let the historian argue the minutiae of terrain and historical placement, but a King Tiger tank in *PanzerBlitz* had better be more powerful than a Russian T-34, and a cavalry charge across a game board should have the same effect as one in real life.

The Assassin

Next we come to the killer gamer, the "assassin." He's the Vince Lombardi type for whom winning is the *only* thing. Balance is important—unless he gets to play the stronger side. In the assassin's view, a game ought to be complex enough to demonstrate his skill, but not so complicated that he has to fight the rules as well as his opponent.

Realism matters only so far as it validates his victories (or, at least, does not invalidate them). When an utter lack of competition forces him to play a game by himself, his aim is to outdo the general of the past—to surpass Napoleon by dealing Wellington a crushing defeat. Within broad limits, he'll play games of any size, scale, and scope. He makes a slashing opponent in chess, a ruthless landlord in the game of *Monopoly*. He goes to conventions to enter tournaments and make new victims of unsuspecting strangers. He plays for blood and considers it an insult to his manhood that women are allowed to play wargames at all.

The Competitor

The assassin's views on complexity and solitaire gaming are shared in large part by the competitor—for which reason the two types are often confused. The competitor, however, seeks challenge more than victory. For him, a wargame is preeminently a contest of skill. He plays chess (or go or bridge, but not *Monopoly*), but prefers wargames because of their excitement, novelty, and absence of fixed lines of play. Balance is a game's most important quality, realism one of the least. He would rather play *Waterloo* than *Panzergruppe Guderian*. He buys fewer games than some others in the hobby, but he plays the ones he owns repeatedly.

The Hobbyist

Of the bunch, the hobbyist is the collector. He subscribes to everything and, if reasonably affluent, may buy a new game every week. He has magazines that saw only one issue and games nobody else ever heard of. He'll buy a game just because its subject is obscure and unusual; yet, in his quest for the perfect wargame, he will collect seven versions of the Battle of Waterloo. He purchases games as much to look at, read, and have as he does to play them. In fact, he spends proportionately fewer of his game-related hours actually playing than anyone else. He's too busy reading, concocting variants and alternative orders of battle, trying to get letters printed in *The General* magazine or articles published in *MOVES*, and designing his own games. For him, the most important aspect of a game is its novelty—in size, scope, scale, subject, or system. He placed an order for *War in the East* as soon as it was announced and bought *Squad Leader* the day it appeared. Along with the historian (the only one who knows more military history), the hobbyist has traditionally constituted the bulk of SPI's target audience. He is the hobby's most vocal and visible representative.

The Gamer

The gamer just likes games. He plays *Clue*, *Diplomacy*, *Dungeons & Dragons*, *Master Mind*, *Ogre*—almost any sort of game, as long as it's a good one. While he may be attracted by the relative realism of wargames, he cares no more about total accuracy than the competitor. Playability's the thing. He plays for enjoyment, for the fun he gets out of interacting with the system and the other players; he doesn't want to spend all day reading rules. With so many games to choose from, he doesn't want to devote his entire life to a single one, either. Long or highly involved games must be correspondingly rewarding. For the gamer, *Diplomacy*, *Dungeons & Dragons*, or *Kingmaker* might

Cover of Avalon Hill's wargame magazine The General. (PHOTO: AVALON HILL GAME CO.)

qualify; *USN* would not. He and the competitor form the traditional Avalon Hill market.

The Specialist

Finally there's the fan, the specialist. For this kind of fanatic, only two things count about a game: subject and quality—with the latter a poor second. Games, in this case, may be only facets of his interest. Someone whose specialty was the eastern front would have volumes on Hitler's war with Russia and miniatures rules as well as board games on the subject. The fan of science fiction is almost certainly a reader—and perhaps a collector—of genre books and magazines; he subscribes to *The Dragon* magazine and *The Space Gamer,* as well as other, more esoteric journals. And, depending on his budget and the seriousness of his addiction, he might acquire Avalon Hill's *Starship Troopers,* SPI's *StarForce Trilogy,* and nearly everything put out by Metagaming, The Chaosium, and Automated Simulations. The most highly rated World War II game wouldn't interest him in the slightest. If he wants a game with tanks, he plays *Ogre* —and gets a supertank. For a "naval" engagement, he might try *Starfleet Orion* on someone's microcomputer. He's fascinated by the possibilities of the future, and gaming allows him to take an active, if limited, role in exploring them. In essence, it allows him to play a character in his favorite kind of story—personally destroying the Klingons, rescuing Dejah Thoris, slipping through the blockade of the oppressive Stel-

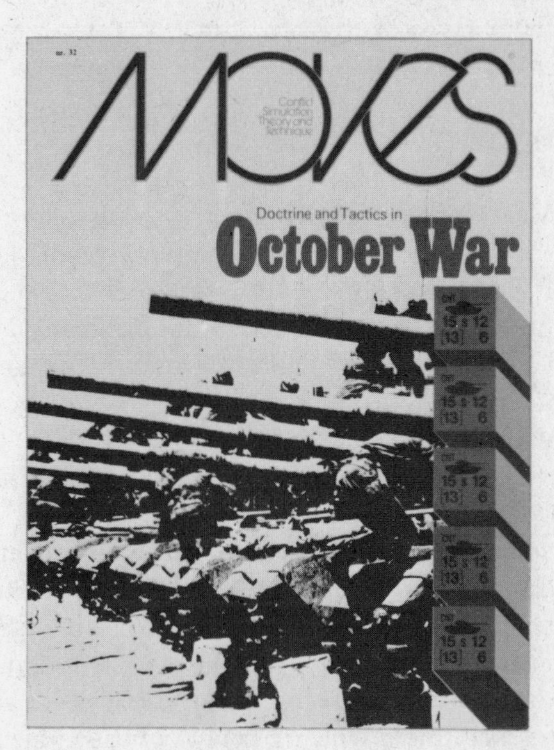

Cover of SPI's magazine MOVES. (PHOTO: SIMULATIONS PUBLICATIONS, INC.)

lar Union, or giving the Bugs what's coming to them—and, perhaps, to rewrite the ending.

Now, like all generalizations, these classifications shouldn't be taken too literally. Few wargamers are perfect examples of the pure genotype. Most are a combination of several varieties: a gamer who's also a science-fiction fan, for instance, or a military enthusiast-assassin type with a special interest in naval warfare in the Age of Sail. Nonetheless, elements of these seven types can be found in just about everyone who plays wargames.

The reasons behind their involvement vary from type to type and individual to individual, but the roots can be traced down to the classic duo: enjoyment and enlightenment. People play wargames for much the same reasons they read *Paradise Lost,* watch television, climb mountains, take flying lessons, or contemplate their navels—to enjoy themselves or to learn something. Or both.

HOW TO GET INVOLVED

Most people are introduced to wargames by someone who is already actively involved. If that was true in your case, you'll have no problem pursuing the matter further: the veteran who got you into it can guide you as far as you like, and he's probably tugging on your sleeve already.

Perhaps, however, you just picked up *The Complete Book of Wargames* or, as one of its authors did many years ago, bought a game on your own, simply because it looked intriguing. In that case, your first step (after finishing the book, of course) is obvious: look over some of the games recommended for beginners, and try a few whose subject matter appeals to you. After that, your choices start multiplying like rabbits.

Buying games and playing them by yourself may be fun for a while, but sooner or later you're going to want to contact other aficionados—to find opponents, exchange information, or simply let the wargaming world know you exist. Converting a friend can provide an immediate, accessible opponent who is of roughly your level of experience and skill. It has the further advantage of providing company and moral support for some of the other steps.

Once you have decided to take the plunge, there are five jumping-off points: clubs, conventions, magazines, the game companies, and hobby/specialty stores.

The store at which you buy wargames is generally a good place to get information. Many of them have bulletin boards full of "opponents wanted" notices and club announcements. The sales personnel are also likely to know of any wargaming clubs or groups in the vicinity. If they don't, or if you buy your games at a department or discount store, try bulletin boards at your local high school or college, or look for ads in school or local newspapers. Failing that, check with a school's history department or a newspaper's feature editor for leads.

Currently, the only thing resembling a national organization for wargamers is AHIKS, formerly the Avalon Hill International Kriegspiel Society, which is devoted to playing wargames by mail. You can get a copy of their magazine, AHIKS *Kommandeur,* by writing the editor at 1111 East Loma Vista Drive, Tempe, Arizona 85282. There are also local groups all over the country. The ones that have retained any particular significance generally have a membership that has outgrown the neo-Nazi

Scene at GenCon, an annual convention of wargamers in Lake Geneva, Wisconsin. (PHOTO: TSR GAMES, INC.)

names and juvenile posturing that characterized many of the early clubs. Probably the best known are Interest Group Baltimore, which has long served as a testing ground for Avalon Hill products, and Metro Detroit Gamers, which sponsors an annual convention (Michicon) and hosted Origins '78.

For a crash course on what the hobby has to offer, nothing beats Origins, the national wargaming convention. The exact date and location of this annual event vary. In 1978 it was held at the University of Michigan; Origins '79 took place on the campus of Widener College, near Philadelphia. Regardless of the location, it lasts a weekend and is attended by thousands of enthusiastic gamers from California to England and from Canada to Florida. Representatives from every game company worth the name are on hand to answer questions, introduce new products, and give demonstrations; and to sell every item in their catalogues, including board games, miniatures, accessories, anything. In more lectures, seminars, and panel discussions than you could possibly attend, game designers explain how your favorite game came to be created; company executives relate the behind-the-scenes story of the wargaming business; successful players reveal their favorite strategies; and a variety of people discuss everything from how to get a club started to why your favorite science-fiction novel didn't get made into a game. There are fifty or a hundred tournaments featuring as many different games and an uncountable number of impromptu sessions of everything from *Black Box, Rail Baron,* and *Cosmic Encounter* to *Panzergruppe Guderian* and Napoleonic miniatures.

The number of other conventions is growing as fast as wargaming itself. These range from purely local one-day affairs to large regional events like Michicon and California's Dundracon and Pacific Encounters. Aside from Origins, the only convention with a real claim to national attendance is Gencon, which is held in or near Lake Geneva, Wisconsin, the home of the sponsoring company, TSR. Not too surprisingly, a *Dungeons & Dragons* tournament is a major feature of the event's program.

While most of the magazines devoted to wargames are put out by gaming companies, several of them, like *Campaign* and *Fire & Movement,* are independent. Most carry reviews of new games, suggestions for new scenarios and variants (different situations or battles using the map and pieces to a published game but with new rules or alternative setups), debates about the merits of a particular system or the historical accuracy of a given game, and articles revealing someone's pet plan of attack. Any of these will widen your gaming horizons in one direction or another. Additionally, *The Space Gamer* carries a convention listing and calendar of upcoming events, and the back

Participants at GenCon, an annual wargame convention in Lake Geneva, Wisconsin, take part in wargames. (PHOTO: TSR GAMES, INC.)

Wargamers battle at GenCon, held annually near Lake Geneva, Wisconsin. (PHOTO: TSR GAMES, INC.)

page of *The General* is full of ads from players seeking face-to-face (ftf) or play-by-mail (pbm) opponents.

As a last resort, try the game companies themselves for answers and advice. For the people who run Avalon Hill, SPI, and the rest, gaming is as much a hobby as a business, and they tend to treat new gamers as fellow enthusiasts as well as potentially valuable customers. If you're polite and patient, they will help you with your problem or question or refer you to someone who can do so.

The world of wargames is varied, exciting, and growing. Come on in and enjoy it!

CHAPTER 2

The Nature of the Beast

WHAT is a wargame? Or simulation? Or whatever it's called?

Although usage varies, in this book "wargame" and "conflict simulation" refer to the same thing: a game whose subject is combat or some similar form of direct conflict between individuals, armies, "nations," or entire species; particularly board wargames of the Charles S. Roberts type, whose mechanics are discussed in detail in the next chapter. Before we get to the nuts and bolts of wargames, however, we'll examine the nature of simulations, their problems and limitations, and their possible subjects and scope.

Most broadly, a simulation is an imitation of reality. For our purposes, a simulation is a replication or re-creation—usually in miniature—of a real or imaginary situation. Simulations, in this general sense, concern us only if they are further characterized by both flexibility and verisimilitude.

To be a game, an "imitation of life" must be flexible. It must allow you to "play" with it. Books, movies, and television fail this test: The Man Who Knows The Secret always gets killed before he can explain his inevitably cryptic telephone call, but the hero, nonetheless, always figures things out in the end, at which point the villain gets shot (in a book), blown up (in a movie), or arrested (on television). We can't relieve the victim's neurotic fear of speaking on the telephone or prevent the villain's last-minute bungling, and we have no way of finding out what would have happened if someone had done something intelligent, like relay the Vital Information to someone else. In varying degrees, however, model railroading, RAND's computer simulations of the economy in the 1990s, and all board games allow us these kinds of options.

Conversely, a game is not a simulation without verisimilitude, the "lifelikeness" that is popularly, if not always correctly, referred to as "realism." The game of checkers is obviously not a re-creation of anything in life; neither is backgammon. The game of *Monopoly*, while having some vague resemblance to the real estate business, is still much too unrealistic to be considered a true simulation. In real life, we don't have to drive down a particular street to buy property there; rent is not fixed permanently and arbitrarily (except perhaps in New York City); our utility bills are not determined by chance; we need not purchase (and tear down) houses before erecting a hotel; and the income tax rate is rather more than ten percent. On the other hand, a simulation of the Battle of Waterloo will attempt to mirror, with substantial accuracy, the forces present and their actual capabilities, the terrain of the battleground and its effect on the conduct of the engagement, and the tactics and aims of the opposing armies. "Luck" will be confined to the historical uncertainties of combat.

This brings us to that old bugaboo: realism versus playability. If this is not the burning issue that it once was, it's only because everyone has become tired of the argument—not because a conclusion was reached.

The question is: How realistic (read: long and complex) can a simulation be before it becomes too long and involved to be played—before it ceases to be a game? Or, how much can be done to make a game easy to play before it ceases to be a simulation? There was a time when Simulations Publications seemed to feel that simultaneous movement—and the tedious order-writing *their* brand of "SiMov" required—and randomly disobedient units were required by realism. There was also a time when Avalon Hill denied that anything with written orders, "command control," or incredibly detailed rules could be playable. After the *Napoleon at War* and *Modern Battles* QuadriGames (on the one hand) and *Squad Leader, Wooden Ships & Iron Men,* and *Tobruk* (on the other), those answers seem as humorously dated as flattop haircuts.

The question has been argued regularly in clubs and magazines and, implicitly, every time a new game is marketed. There is no single answer, for the same reason that there is no single answer to why people play wargames. The idiosyncratic individuals who play wargames for reasons of their own are free to argue the matter endlessly, and to buy those games that suit their individual notions of realism and playability.

Still, a simulation game ought to meet certain requirements. It should be internally consistent. This may be most readily explained through some gross examples: if one unit is allowed to move four hexes every turn, another ought not to move according to the roll of the die. If an artillery piece attacks by comparing its "attack strength" to an opposing piece's "defense strength," it would be objectionable to have an infantry unit capture by replacement (as in chess) and a cavalry unit take another by jumping over it (as in checkers). Less absurdly, perhaps, a rule ought not to have more exceptions than cases in which it applies. A number, counter, color, or term shouldn't mean one thing in one place and something quite different somewhere else. And so on.

These instances are not as exaggerated as they might seem. The role-playing games discussed elsewhere are notorious for faults of this kind. In *Dungeons & Dragons,* for example, the term "level" is used in four distinct ways, none of which is explicitly defined! The numbers used to characterize "armor classes"—which run, illogically and perversely, backward—are utterly unrelated (or, at best/worst, *inversely* related) to any other numbers, die rolls, or systems used in the game. The newer editions of *Tunnels & Trolls* admit that, if poisoned weapons are used, you pretty much have to throw out the game's entire system of combat!

Nor are standard wargames exempt. The rules of *Outreach,* for example, have, practically on top of one another, two sets of two paragraphs that directly contradict each other. In other games there are references to imaginary paragraphs, promised explanations that are never given, and rules for units and situations that don't exist. The infamous *Fall of Rome* elevated such contradictions to a high art.

Simply put, a game is supposed to make sense. Too often it does not. Much of the complication in modern wargames is due not to depth but to too many special cases, exceptions, and one-time-only rules. Perhaps because of the hectic pace of current production, game designers don't always take the time to come up with truly elegant rules: those that can be stated clearly and simply but remain of general applicability.

A historical or contemporary simulation, moreover, must reflect a reality outside the game. This means that battleships can't travel faster than aircraft, that Robert E. Lee did not command the Army of the Potomac in the Civil War, that a rifleman can't stop a tank by himself, and that a cavalry charge against English longbowmen or a Macedonian phalanx is not going to be wildly successful.

A game must represent the possibilities of the engagement. Had things gone just a bit differently, could Napoleon have won at Waterloo or Caesar have lost at Alesia? Probably. Those outcomes, therefore, must be possible in games based on those battles. Yet, you can't have a conflict simulation in which Custer defeats the Sioux at the Little Big Horn or the German battleship *Bismarck* sinks the entire British Navy. These things simply could not have happened. But *relative* victory—measured in endurance time or enemy casualties—is another story.

A simulation should allow the actual historical outcome, but in fact this is often a big

Selection of board wargames available from Heritage Models. (PHOTO: HERITAGE MODELS, INC.)

problem. There is no good way to simulate surprise or ignorance. Consider the Trojan horse. If you were the Trojan player, would you allow the thing inside the "topless towers of Ilium," knowing Cassandra was right and Troy was doomed if you did? Of course not—not unless the matter was taken out of your hands and left entirely up to a die roll, as it is in *Troy*. How satisfactory is that?

Similarly, the success of the German blitzkrieg in France in World War II was due to surprise and the lack of French preparedness. *France, 1940* was designed to re-create not just the German victory but its magnitude as well; it couldn't simulate ignorance, but it did, quite cleverly, prevent the French player's knowledge from helping him. Unfortunately, playing the French against an unstoppable (if contrived) German juggernaut was as much fun as observing the action of a steamroller as it prepares to pass over you.

You can't go back and repeat the surprises of history, because they aren't surprises the second time around. And you can't spring real surprises of your own, because the rules are not broad enough to encompass the universe of improbable actions. A good role-playing game does allow this freedom of action, but such games don't attempt to re-create the battles of history. A computer can be programmed to fall for the same trick every time, but if that's the only way you can win, how much fun is it? And how much of a contest?

This basic constraint, however, doesn't prevent a simulation from being successful. It can still be an interesting, challenging contest of tactics and strategy. If you want to repeat a particular historical ploy, solitaire play allows you to make either side as inept as you wish. But don't expect a live opponent to play a dumfounded Dr. Watson to your brilliant Holmes, especially if he's read the book.

FROM MIDWAY TO MIDDLE EARTH

From *Stalingrad, Gettysburg,* and *Waterloo* to *Wacht am Rhein, Terrible Swift Sword,* and *Air Assault on Crete,* the vast majority of contemporary wargames have been based on actual battles. Since most such conflicts have been two-sided, most wargames are, at least nominally, designed for two players. In effect, one player takes the part of, say, Napoleon and his officers; he maneuvers the pieces representing the French forces; and he tries to destroy or drive off the Allied pieces and advance on Brussels. His opponent acts as the Duke of Wellington and the Allied officers: using pieces that symbolize the English and Prussian soldiers present at Waterloo, he tries to equal Wellington's historical victory over the French of Napoleon.

Depending on the scale used in the game, the playing surface may be small or so large it requires several boards, and there may be a dozen or hundreds of pieces on a side. As long as the pieces portray with reasonable accuracy the relative abilities of the forces, with similar outcomes under similar circumstances, one scale is as realistic as another.

In a hopelessly narrow sense, a game stops simulating the battle as it was actually fought as soon as a single piece is moved to a location never occupied by the real army/regiment/platoon to which it corresponds. Such a strict definition, however, makes a simulation *game* an impossibility, since a game without options—like poker played with a stacked deck—isn't really a game at all.

It is more useful, therefore, to distinguish games that re-create the terrain, forces, aims, and initial positions of a historical conflict from those versions or variations that introduce major alternatives in the form of a revised order of battle (the forces present), the timing of related events (such as the arrival of reinforcements), or major changes in location, preparedness, knowledge, or policies.

An example or two may be helpful. The last day of the Battle of Waterloo might have gone very differently had Grouchy not spent the previous day wandering all over the countryside looking for the defeated but regrouping Prussian Army. An obvious alternative, then, would be to refight the day of June 18, under the assumption that Blücher's Prussians were not available to reinforce Wellington's army. Similarly, Hitler's eastern-front campaign would not have been such a complete disaster for the Third Reich had he not promulgated his notorious "no retreat" order.

Besides allowing us to satisfy a bump of curiosity, such semihistorical "what-ifs" and "might-have-beens" can also serve to balance some of what were, historically, rather one-sided contests. Even the unbalanced *France, 1940* included several alternate orders of battle—reflecting, in theory, a difference in policy, prewar buildup, or British assistance—aimed at giving the French player more of a chance.

Such alternatives are a common feature of contemporary wargames, but they rarely are packaged separately as games in themselves. A notable exception is *D-Day*, in which the Allied player is free to invade any one of seven different coastal areas, only one of which (Normandy, of course) was actually assaulted on June 6, 1944. A basic part of the game's appeal is the chance to explore the consequences of alternate invasion sites.

Still farther away from history-as-it-happened are games that allow, or are based on, changes so great that the opposing countries are different. In Avalon Hill's *Third Reich*, for instance, the German player is given the option of not fighting Russia at all —a move that obviously would change World War II immeasurably. *Diplomacy*, which simulates the flavor of turn-of-the-century diplomacy rather than its factual details, allows such unlikely occurrences as an Anglo-German attack on France.

More typical are games based entirely on battles that never took place but that slightly different circumstances might have made plausible. There are several games based on the proposed Operation Sea Lion, an invasion of England seriously considered by the German High Command in World War II. SPI's *Operation Olympic*, which is no longer available, was a solitaire game dealing with a possible Allied invasion of Japan in the same war. The subject of a more recent game, Avalon Hill's *Invasion of Malta —1942*, is equally hypothetical.

Despite what may be initial impressions to the contrary, games like these do not lack historical value. By experiencing firsthand, as it were, the difficulties of an overseas assault, a player in SPI's *Seelowe* can get a better understanding of why Nazi Germany decided not to put Sea Lion into operation.

In addition to curiosity and novelty value, a game based on a battle that never took place has two major attractions for nonhistorians. First, you are not playing in the shadow of Napoleon or Rommel; that is, there is no built-in critical standard by which your "generalship" must be judged. Second, you do not have the original (real) tactics and strategies to fall back on: you must be innovative.

These same reasons help to explain the popularity of the hypothetical-contemporary games (as well as the straight science-fiction and fantasy games), many of which are based on battles that have not taken place but that might occur: *Mech War '77*, the new and old versions of *NATO* and *Red Star/White Star*, and the two *Modern Battles* QuadriGames, all of which, at least in part, are based on a Warsaw Pact invasion of Germany; *The East Is Red*, which simulates a Sino-Soviet war; *Oil War*, a fight for Middle East oil reserves; *World War Three; Air War;* and others. Broadly speaking, all these could be considered science fiction, but they can be distinguished from *Ogre* or *After the Holocaust* by their strong historical flavor. The element of speculation is quite limited: Although open to interpretation and subject to certain "Classified" gaps, the data for such potential conflicts are available today. Games like these may not interest the pure "historian," but, if the research behind them is accurate, they cannot be faulted on grounds of realism: tomorrow's sunrise can't be said to be any less realistic than yesterday's simply because it hasn't happened yet.

While many gamers like to dwell on the events of the past, others prefer to speculate about the unforeseeable but imaginable fu-

Simulations Publications' World War III. (PHOTO: SIMULATIONS PUBLICATIONS, INC.)

ture or the imaginary and improbable past. As models, science-fiction and fantasy games may take the logically extrapolated near future as it might be (*Invasion: America*), the present as it might have been had events of the past turned out differently (*Dixie*), the secondary world of Tolkien's Middle Earth (*War of the Ring*), the created future of Poul Anderson's Polesotechnic League (*The Ythri*), or the stories of Michael Moorcock's pyrotechnic if undisciplined imagination. Alternatively, an elaborate and original future history may be developed, as SPI has done with its *StarForce* trilogy and separately with *BattleFleet Mars,* Automated Simulations with *Starfleet Orion* and *Invasion Orion,* and The Chaosium with a variety of games set on the world of Glorantha. Science-fiction and fantasy games may not appear to be "simulations" at all, but even those with the sketchiest of backgrounds all attempt to evoke—for the duration of play, at least—a separate reality.

Directly challenging a player's intuitive insights, games with extraordinary subject matter and novel lines of play test the flexibility of his intelligence and the limits of his imagination. Unhampered by the fetters of the past, players of these games must bring to the contest only the weapons of a karate master: open hands, a clear mind, and a willing spirit.

AND THE BREADTH OF IT FIFTY CUBITS

Wargames vary as much in size and scope as they do in period and setting.

At one end of the scale are science-fiction games that treat not just the fate of a planet but the future of the human race—and others—on dozens of worlds in as many star systems. On a map representing tens of thousands of light-years, *Outreach* simulates the expansion of several groups of intelligent species from the neighborhood of our spiral arm to the center of the galaxy. The game *4000 A.D.* confines itself to a mere forty-eight star systems in our stellar vicinity. *Star Web* and *Steller Conquest* both use imaginary configurations of stars and planets. All four games deal with grand strategy—the big picture. Although the details vary from game to game, players must deal with the enormous problems of exploration, expansion, population growth, resource exploitation, commerce, and shipbuilding. You must not only decide how to wage a war, but also when, where, whether, and even *who* to fight. Unlike the vast majority of wargames, this group is designed for more than two players.

A step down is the simulation of a complete war. In science-fiction games like *Imperium,* this may involve several worlds, but more typical are *World War II* and *Third Reich.* Although neither of the latter two games deals with the Pacific theater (virtually a war in itself), both contain elements of grand strategy. Other similar games, like *1776,* may be more strictly strategic in scope. The difference is one of options. In this sense, strategy deals with questions like where to invade, with what kind of troops, and how much effort to expend in which countries or regions. Grand strategy subsumes all this and more: Do you ally with Russia (or Alpha Centauri), remain neutral, or attack? Do you want to expend your resources building fleets or armies? Aircraft or tanks? Fighters or bombers?

Units in games of this scope are large groups—whole armies, or at least corps-level organizations. Combat is fairly abstract and based on the ratio of the strengths of the opposing forces in any particular battle.

A war can be subdivided into phases or geographic areas. Depending on the historical period, it might be appropriate to simulate a campaign in, say, the Civil War or a front in either of the world wars. *D-Day* is the classic example of the latter, but *Atlantic Wall*, a newer treatment of the same subject, is more typical of recent designs, which include the longest-playing, largest, most complex, and most notorious games in the hobby. The pieces in *War in the East, War in the West, Drang Nach Osten!*, and *Unentschieden* represent divisions or brigades—units that are unexpectedly smaller than those that typify the simulations of a complete war. As a result, there are more than a thousand pieces in each game, and the unusual scale requires up to half a dozen mapboards and a playing area no smaller than that called for by many miniatures campaigns. Save for pride of ownership, games like these are suitable chiefly for clubs and essays on "What I Did on My Summer Vacation." SPI's term "operational" is sometimes used to describe both this class (as distinguished from a "strategic" version of a similar subject, like *Russian Campaign*, which uses army- or corps-level units) and those games (*Panzergruppe Guderian*) that simulate a smaller portion of a war—a campaign or an "operation."

The most obvious grouping contains those games that treat a single battle. Depending on the extent of the actual conflict and the scale used to simulate it, it's possible to have a strategic, operational, or grand tactical treatment of a single subject. (*Gettysburg '77*, Avalon Hill's newest version of its oldest game, is one of the few that can claim to include all three approaches in a single package.) The term "grand tactical" is often used, generally, to describe all "battle" games, particularly for engagements prior to the twentieth century, or to refer, specifically, to those using a scaled-down *War in the East* approach. *La Bataille de la Moscowa, Terrible Swift Sword*, and *Wellington's Victory* all utilize huge mapboards, elaborate rules, and battalion-to-regiment-size units to simulate in loving, if possibly bewildering, detail the battles of Borodino, Gettysburg, and Waterloo, respectively.

Tactical games of whatever scale (from *grand* to *close*) usually have less abstract combat systems than strategic games. Ranged fire, which allows a unit to attack or "shoot at" an opponent's piece from a distance, is typical not just of artillery units, as in some operational games, but of all, or nearly all, units: ships, planes, tanks, infantry. This allows a more vivid simulation that many gamers find more exciting. Despite the complexities caused by the detail involved, newcomers may actually find it easier to grasp tactical games—particularly if they are sea or air games and few pieces are involved.

Many tactical games simulate a form or period of warfare rather than a particular engagement. Most include a multitude of scenarios (minigames with essentially the same rules but different setups and objectives) that may represent actual battles or merely typical actions. Pieces normally represent individual ships and planes or small units of armor and infantry (platoons or companies). *PanzerBlitz*, the father of modern tactical games, at least on land, covers armored warfare on the eastern front; *Panzer Leader*, as much son as brother, does the same for the western front. *Frigate* and *Wooden*

Ships & Iron Men deal with naval warfare in the Age of Sail; *CA* and *Dreadnought* perform the same service for this century. *Richthofen's War* simulates air combat in World War I; *Air War* and the older *Foxbat & Phantom* handle contemporary aircraft.

At the small end of the scale, the tactical binge of the past decade has culminated in a handful of close tactical games. Pieces in *Sniper!* and *Patrol* represent single men, and the objectives include clearing a building of enemy snipers or doing the same to a section of forest. Although the activity scale was immediately popular, the simultaneous movement system requires written "plots" or orders for every action taken by every piece. *Squad Leader*—along with its successor, *Cross of Iron*, the newest sensation along this line—uses a slightly larger scale: units may be infantry squads or individual tanks, machine guns or leaders. This latter system allows a greater variety of forces than *Sniper!* and, through the use of intricately interwoven movement and fire sequences, gives something of the feel of simultaneity while avoiding the tedium of order-writing.

Differences in scale are as great as those between separate periods of history. There are people who play nothing but tactical games just as there are those who sneer at any subject set before World War II. There is, nonetheless, no ideal scale any more than there could be a "perfect" war. It depends on what you want to simulate and how big or complicated you want a game to be. With hundreds of wargames on the market, you have plenty of choice.

CHAPTER 3

All's Not Fair: A Concise Guide to Playing Wargames

DESPITE THE great variety of wargames on the market, most of them have a lot in common. This is natural and necessary; otherwise, a new game would require a much longer learning period than is usually the case. While each game has some unique aspect, and while there are exceptions to every rule, these common elements allow us to talk about wargames and how they work in general terms.

Like any other hobby or sport, wargaming has a special language. Unlike government bureaucratese, wargame jargon is not essentially obfuscatory: it's intended to enlighten, not to confuse. Often, its stylistic abuses stem from a misguided attempt at precision or a simple lack of writing skill.

In this book, terms that may be unfamiliar to you are normally defined the first time or two they are used. If you don't know or can't remember the meaning of a word or phrase, and it isn't clear from the context, check the Glossary.

In addition to a common language, most wargames have a number of parts in common: counters or pieces, a board, a die or two, some charts and tables, and, of course, a folder or booklet of rules.

THE BOARD

While miniatures campaigns and role-playing games use tables or replica terrain, standard wargames involve one or more mapboards. Some, like those made by the Avalon Hill Company, are mounted; the printed map on which the pieces are moved is glued to heavy cardboard or fiberboard. Unmounted boards or mapsheets are simply large pieces of heavy paper printed with appropriate details. Simulations Publications and many smaller publishers supply unmounted boards with most of their games. Although most gamers prefer one or the other, there are advantages to both.

A completely blank game board would be meaningless. Something must separate one area or space from another, or the players could not tell how far they were moving or where they were doing anything. A wargame board is divided into sections in one of three ways. *Diplomacy*, *War at Sea*, and a few other games, like the family game *Risk*, divide their boards into geographical areas of no particular shape: Spain might be one area, the Burgundy region of France another, and the North Sea a third. Early wargames like *Tactics II* used squares like a chess or go board or a piece of graph paper. They resembled city road maps that are divided into squares to help you locate streets. Most of the games covered in this book, however, have boards that are divided into small hexagons—"hexes," for short.

Regardless of their shape, these divisions serve the same purposes as the squares on a chess board: the differentiation of placement and function (ships or fleets go in the blue—water—area; armies or tanks go on the brown spaces). The divisions also regulate movement and combat (this infantry unit can move four hexes per turn; that artillery unit has a range—in combat—of six hexes). Hexes, fairly obviously, allow finer distinctions and greater precision than irregularly shaped areas.

In certain air, space, or naval games the boards are, except for the hexagonal grid, blank or uniformly colored or patterned. In such cases, the medium being simulated has no significant terrain features that affect play. For example, clouds don't slow down supersonic aircraft. In most cases, however, the board also will be imprinted with something resembling a map. Different colors, patterns, markings, or lettering may distinguish land, sea, swamp, forest, plains, a range of hills or mountains, rivers, roads, railways, towns, fortifications (concrete bunkers, trenches, fallout shelters, whatever)—any sort of terrain that might affect

Tactics II, *an early wargame published by Avalon Hill.* (PHOTO: AVALON HILL GAME CO.)

movement, combat, or other game functions.

This map may represent anything from a small portion of a single town (*Sniper!*) to a large part of our galaxy (*Outreach*). Some tactical games, like *PanzerBlitz* and *Squad Leader*, use several small boards, each of which depicts typical terrain rather than representing an actual location. These "isomorphic" boards can be placed side to side or end to end in a variety of configurations, depending on the game scenario being played and the locale being simulated.

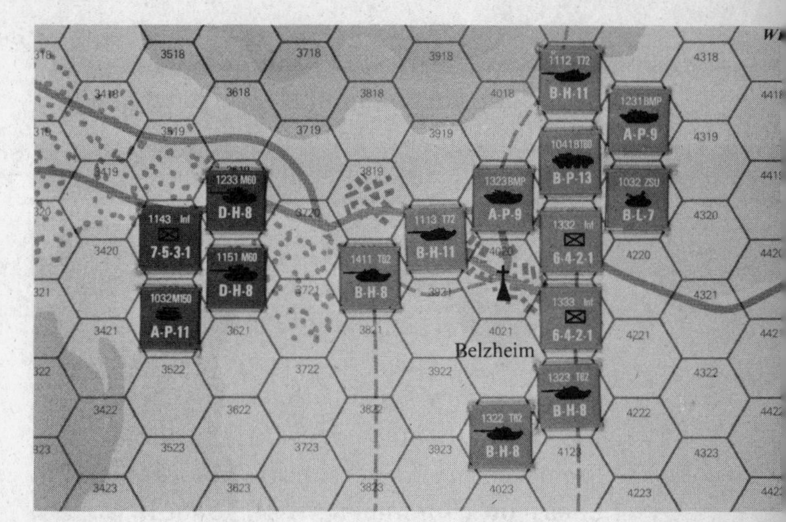

Counters in position on a game mapboard.
(PHOTO: SIMULATIONS PUBLICATIONS, INC.)

Hexes usually range in size from a half inch to an inch across. The relationship between the size of the hexes and the amount of territory represented governs—and is governed by—the level or scale of the game. If each hex represented a distance of only six feet, the game would be on a close tactical level; the pieces or units would probably be individuals, and a turn would correspond to a few seconds. On the other end of the scale, hexes simulating 1,500 light-years would indicate a strategic or grand strategic-level game, in which a turn might stand for a generation or more.

The network of terrain hexes also designates the basic situation of the game. In *D-Day*, for instance, the Allies must initially place their troops adjacent to beach hexes in just one of seven possible invasion areas. Their immediate tactical goal is to land successfully—to drive the German units back and establish a beachhead that will allow reinforcements to land. The German player's immediate tactical goal is to prevent Allied units from accomplishing this or, at least, to make it as slow, difficult, and costly as possible. In the long run, the Allies' strategic goal is to march overland and cross the Rhine River with enough units to secure a position there.

In *Kassala*, an introductory game for you to try in the next chapter, the situation can also be defined by the terrain. The Christian troops have strong defensive positions inside the towns of Udaka and Kassala. To win, the Moslem forces must drive the Christians out of the town hexes. Terrain further influences tactical decisions: the Christians would be foolish to abandon their strong positions in the towns and behind the trenches and, instead, to charge into the open against the numerically superior Moslem forces. By the same token, a river (on the east) and wadi hex-sides (on the west) prevent the Moslem cavalry from outflanking the defenders.

A CARDBOARD ARMY

Pieces or tokens, such as plastic pawns, wooden checkers, or metal racing cars, are common to many games. *Nieuchess*, an early quasi-wargame from Avalon Hill, used two groups of common plastic pawns to represent opposing armies, but this simple measure is not suitable for contemporary wargames. In ordinary games, the only requirement is a means of distinguishing one player's pieces from another's. The obvious answer—different colors—works well whether a person has only one token or a whole group, as in checkers. Things get more complicated when further distinctions are necessary. Differentiation by shape, as in chess or *Diplomacy*, will go only so far: there are only six different kinds of pieces in chess, but *PanzerBlitz* has dozens of different kinds of armor units alone.

The answer is the cardboard counter. This is a piece of heavy cardboard of varying thickness. Usually it is a half-inch square, but sometimes it is somewhat larger, as in *PanzerBlitz*, and occasionally rectangular like the ship counters used in *Jutland* or *Wooden Ships & Iron Men*. Counters generally come in a group (a counter sheet), die-cut or perforated, much like a sheet of stamps. To reduce cost, some companies use thin counters that are cut only along their vertical edges. The gamer must cut the other edges with scissors or a knife.

Color is used to distinguish the player or side controlling the pieces, but particular shades may also distinguish pieces by national subgroup or function. For example, shades may separate cavalry from infantry or troops in general from markers used as bookkeeping devices.

More important, the counter is imprinted with a central symbol and a series of letters and/or numbers (Fig. 3–1). The symbol identifies the unit by general type, such as cavalry. It may be a silhouette, as in *PanzerBlitz*, or something as abstract as a pair of intersecting lines inside a square, as in *Kassala*. Small letters or numbers around the counter's central symbol detail the size of the unit and its place in the organizational hierarchy: for example, the First Battalion of the Third Brigade. And a word or abbreviation may further identify the unit by nationality, specific type, name, nickname, or commanding officer.

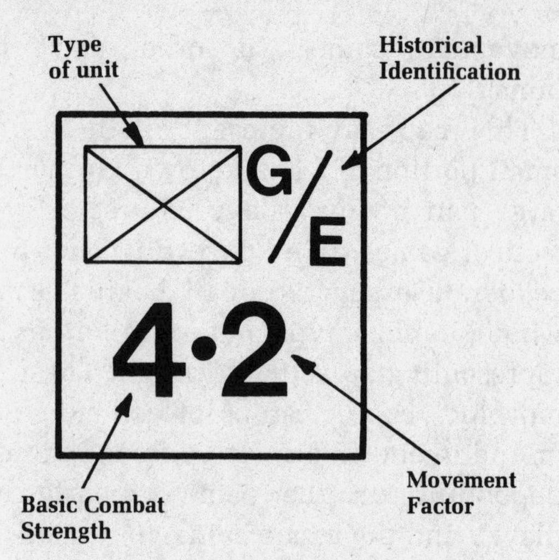

Anatomy of a playing unit or counter.

The other numbers on counters are factors, which are quantifications of the unit's basic abilities in relation to other units and the hex grid on the playing board. A unit with a combat factor of 4 has twice the fighting strength of a unit whose combat factor is only 2. A unit with a movement factor of 4 can be moved twice as far (four spaces or hexes) in a turn as a piece with a movement factor of 2. Games with simple counters generally have only these two factors. Nearly always, the combat factor is the big number on the lower left of the counter, and the movement factor is the big number on the lower right. Other games separate the functions of the combat factor into attack and defense factors. Still other tactical games may also have a range indication. This tells how far, in hexes, a unit can be from an

Components of Avalon Hill's World War II game Squad Leader. (PHOTO: AVALON HILL GAME CO.)

enemy and still attack it. Additionally, there may be a morale factor, which indicates how likely it is that a unit will panic, retreat, or disobey orders.

The chief virtues of this system are that it is compact and extremely flexible, but since the meaning and location of these numbers vary from game to game, it may seem confusing at first. All wargames, however, have a picture, chart, or written explanation in the rules that explains what factors are represented on counters and where. With very little practice, you can distinguish a 3·3 Marmaluke cavalry unit from a 4·2 Portuguese infantry unit as easily as you can tell a queen from a knight in chess.

Some recent games, meanwhile, feature back-printed counters that are color-coded on both sides. This can effectively double the number of units from which to choose, but it usually serves other purposes.

If the counter is colored on the back, but is otherwise blank, it will probably be placed on the board and, at least initially, moved upside-down. This keeps the opposing player from knowing what units are where and, particularly if some of the pieces are dummy counters (blank on both sides), it deprives him of his unrealistic edge over the generals of history, few of whom ever had total knowledge of the disposition and movements of their enemy's forces. When these inverted units come into contact with or, in some games, in sight of enemy units, they are flipped over, and their true nature is revealed.

Sometimes the back side of the counter is identical to the front except for its combat strength (the combat factor). This allows a simple method of "step reduction"; instead of being removed from the board, a defeated unit is flipped over and thereafter fights at, say, half strength (half its component troops having presumably been killed, wounded, disabled, or captured). In the simplest case, a second defeat would eliminate the half-strength unit, although in some games this reduction process might continue for some time by means of substitute counters of diminishing strength.

Back-printing is also used in a system of "untried units." In *Invasion: America* and *Panzergruppe Guderian*, for example, certain counters are printed with two sets of combat factors—a nominal value and a real value. These represent untested troops: those who haven't shown their worth in combat. All such units are placed and moved with the "nominal" side up; when they engage in combat for the first time, they are flipped over to reveal their true combat factors, which may be better or worse (sometimes much worse) than their presumed factors. Like the inverted counters, the purpose of untried units is to give the player a taste of the uncertainties faced by an actual commander.

THE RULES OF THE GAME

It is fairly obvious that the rules vary somewhat from wargame to wargame, and anything with twenty or thirty pages of rules may look far too complicated to assimilate. It's not that bad, though. Ninety-nine percent of wargaming rules break down into four simple categories: movement, combat, victory conditions, and basic specifications.

Let's examine movement first. The basic sequence is: I move—you move—I move again—you move again. It's just like a game of chess or backgammon. The biggest difference is that in wargames you don't have to limit yourself to moving only one or two pieces at a time. Instead, I move all my pieces or as many of them as I wish—you move all your pieces—I move all my pieces again, and so on.

Complication number two is: How far do I move my pieces? This is where the movement factor comes into play. Essentially, each piece can be moved (each turn, remember) as far, in hexes, as its movement factor. An infantry unit in *Kassala* with a movement factor of 2 can move a maximum of two hexes in a turn; it could also move only one hex or none at all. Different units often have different movement factors and, consequently, can move farther than, or not as far as, other units.

Admittedly, it's not usually quite so simple; terrain can modify the distance a unit can move. This is only reasonable. One doesn't slog through a swamp or climb a mountain at the same speed as walking across an open field. This is reflected in what is referred to as an increased cost in movement points to enter, say, a swamp hex or cross a river hex-side. A movement point is simply an increment of the movement factor; a movement factor of 4 gives a unit four movement points. In *Kassala*, it costs an infantry unit an extra movement point to cross a wadi hex-side. In some games such terrain costs are minimal; in others there are many different kinds of terrain, each of which may affect some types of units differently from others. These effects will often be summarized in a Terrain Effects Chart or Key, printed in the rule book or on the board itself.

In most games, that's really all there is to movement. In some, though, there are other complications. In a few games, certain units such as tanks get to move twice during a single turn: once (with everyone else) before combat and once after. This is usually called "double-impulse" or "mechanized" movement. Its purpose is to simulate the breakthroughs typical of armored warfare in World War II.

Finally, there's simultaneous movement. Instead of taking turns, both players move at the same time. To eliminate confusion and to make sure that neither player waits to see what his opponent is doing before committing himself, nearly all simultaneous movement systems require some form of written orders. Sometimes this is referred to as a "movement plot." On a piece of paper, one player writes down where he wants each of his units to move or what he wants them to do; the other player does the same on another piece of paper. When both players finish, they move their units according to what they wrote down. In some games, these orders are read aloud and checked immediately; in others, the records are kept separate until the end of the game. While this system has advantages, it gets extremely tedious if many units are involved or if the orders are complicated.

Wargames, of course, are not simply races. Units do more than just move; they also fight. In wargames, one piece doesn't "take" another just because it moves into the same space, as in chess or backgammon; obviously, that's not realistic. In the simplest case, opposing units fight when one moves next to the other. While not as simple as a capture in chess, the result of combat is not really as complicated as it might first appear.

Basically, the separate combat factors of the units involved in a particular battle are totaled and compared. This comparison is expressed as a fraction or a ratio and simplified so that (in most cases) one of the numbers is 1 (one): for example, 2–1 (read: two-to-one) or 1–3. The "attacking" units—those belonging to the player whose turn it is—are always represented by the first number; the second number always belongs to the defender. If "rounding" is necessary to make the numbers come out even, fractions are nearly always rounded in favor of the defender. For instance, if a Portuguese infantry unit (combat factor 4) attacked a Marmaluke cavalry unit (combat factor 3), the true odds would be 4 to 3, but this would be simplified to 1–1. This final odds ratio, which must always match one of the headings on the Combat Results Table (CRT), is used to determine which column of the CRT to consult. Normally, a die is then rolled, though in some cases a set of numbered counters or "chits" serves the same purpose. The number shown on the die is cross-indexed with the appropriate column, and the result of the battle is then taken from that place on the CRT.

There are, as you might suppose, more complications. Terrain may affect the strength of the attacker or the defender—or both. In *Kassala*, a defending unit in the town of Udaka has its combat strength increased by one. That is, its combat factor is temporarily increased by one increment—a strength point—as from 4 to 5, while a unit attacking from the "wrong" side of a wadi or trench hex-side temporarily loses a strength point. In some games, a town or woods hex might double a defending unit's combat strength, while other terrain might halve an attack.

In many tactical games, attacks may be

made at a distance. A unit's range factor determines how far away it can be and still fight. The German battleship *Bismarck*, you may recall, sunk the British battle cruiser H.M.S. *Hood* from fifteen miles away. Ranged fire is typically subject to line-of-sight (LOS) restrictions: a unit may have to be able to "see" what it's firing at, and a hill hex might obstruct the view. But such combat is resolved much the same as any other.

The basic pattern from which all variations derive is this: I move—I attack/we fight (end of my turn)—you move—you attack/we fight (end of your turn and the game turn). And the cycle is repeated. Variations are endless. If movement is simultaneous, combat may well be too, and if there's double-impulse movement, there may be a second combat phase to go with it. In a single player's turn, there may be a dozen segments, phases, and subphases.

An important point in any game is: What does it take to win? This is explained in a game's victory conditions, which are no more constant than anything else in wargaming. Generally, they divide into two broad categories: material and territorial. In the first case, the object of the game may be to eliminate all your opponent's pieces or perhaps just more than you lose. In the second, the aim is the occupation of particular territory—certain specified hexes—either for, say, four turns in a row or at the end of the game. *Kassala* has such territorial objectives: the occupation of Udaka and Kassala after ten turns. In more complicated cases, which may involve a combination of material and territorial aims, the winner of the game may be determined on a point basis. For example, a player gets one point for every one of his units still on the board at the end of the game and ten points if he occupies the town; the other player gets two points for every one of his opponent's pieces he eliminates and five points if he can keep his opponent out of the town. The player with the higher total of victory points wins the game.

Somewhere in the rules to all wargames will be a list of the pieces or counters that go to each player; often, these forces will be less than the total included with the game. This allows for replacement of lost pieces, reinforcements, or variations and alternatives. If one player is badly beaten in one game, he may get a couple of extra units next time. In *Kassala*, every unit has a fixed starting position printed on the mapboard; in other cases, there may be some freedom of choice in placement. If some units start off the board, the rules will specify when they can be placed on the board, where, and how many in one turn. Sometimes a player is allowed to pick whatever units he wants up to a total of, say, twenty strength points (combat factors). For example, he may select two 5·5's, a 4·4, and a pair of 3·4's.

Finally, many games are actually several games in one. Each of the component games is called a "scenario." While using the same board and rules with perhaps some variation, each scenario posits a slightly different situation, with different forces on each side and different victory conditions. These scenarios are often graded in complexity, so that you can play a relatively simple game with few units while you're familiarizing yourself with the rules. Later, after you're more comfortable with the procedures involved, you can tackle a more complicated situation. A broad selection of scenarios may allow you to pick one that is as long, as complex, as balanced, and designed for as many players as you wish.

THE HAMMER AND THE ANVIL

Dice are used so commonly to control movement in many games, such as backgammon, that many people don't realize other methods are possible. A die (usually only one) is also a wargaming staple. However, since units are moved according to their movement factors, the die in wargames serves a different function: it is the hammer of war.

If we decide to resolve a battle between our pieces by rolling a die, ignoring the strength, size, and number of units involved as well as the presence of fortifications or defensible terrain, that's luck: whoever rolls the highest number gets to take away all the loser's pieces. If, however, we say that one die is rolled for each unit involved, and every 6 rolled eliminates an enemy piece, that is the beginning of a Combat Results Table.

If the die is the hammer of chance, the Combat Results Table is the anvil of probability. Very simply, the CRT is a collection of the results or outcomes of battles fought at each of a specified set of odds—for example, 1–3, 1–2, 1–1, 2–1, 3–1, 4–1, 5–1. Alternatively, the results will be based not on the ratio of the strengths of the opposing forces—a process of division—but on the difference between their adjusted combat strengths—a process of subtraction. Generally, there are six outcomes in each CRT column—one for each number on a die—but not all need differ.

After the combat strengths of the embattled units have been calculated and adjustments made for terrain, fortifications, and possibly even the charisma of the opposing generals, the process of quantification is simplified into a single number or pair of numbers (the odds ratio) that corresponds to one of the column headings on the CRT. Then the die is rolled, and the hammer strikes. It may flatten the combatants, eliminating one side and halving the other. It might rebound harmlessly, leaving the situation essentially unresolved. In effect, the die roll simulates the fluctuations of events caused by surprise, bravery, cowardice, fatigue, luck—the fortunes of war.

The CRT insures that the results are not merely haphazard by altering both the outcomes that can occur in any particular column and the odds that any given outcome will occur. For example, in a battle fought in *Kassala* at the poor odds of 1–4, the chances are two to one that the reckless attacker will be eliminated without cost to the defender. At the far better odds of 2–1, there is only one chance in six that the same thing will happen. And, at 3–1 odds, the best the defender can hope for is a standoff.

The most common type of Combat Results Table gives a single result, which may affect the attacker, the defender, or both, for each odds column and roll of the die. Tactical CRTs for games using ranged fire only affect the defender; they are in effect the results of an attack: one side of a battle rather than the whole thing. Complicated CRTs may have die-roll modifications under certain circumstances and may give two different results, one for the attacker and a separate one for the defender, for each roll. For example, the attacker loses a unit, and the defender retreats two spaces. Although the results in a CRT vary widely from game to game, all have in common a probabilistic structure: the stronger the attack (compared with the defense), the greater the chance for victory.

Modifications to combat caused by the effects of terrain or differences in weapon types are nearly always incorporated into charts or tables like the Terrain Effects Chart. In a well presented game, these charts will be printed on a corner of the board or on a card or sheet that can be left in view for reference. All but the most haphazardly prepared rule books will at least have them placed together, so that players need not hunt through twenty pages of rules every time a different chart is needed.

CUTTING THE GORDIAN KNOT

A thirty-page rule book can be pretty confusing—even frightening. At first glance it looks so awesome that you may despair of ever understanding it. A wargame and its rules, however, are not as complicated as they appear. We have already seen that, despite differences in detail, most wargames share a language and procedure. Furthermore, the biggest part of a standard rule book can be ignored completely, or at least read with less attention than you must give to other parts. Games are often compared with novels, but it may be helpful to consider a game's rule book less like a novel and more like a collection of short stories.

A third of a large rule book might be scenario listings with an explanation of each scenario, background material, a listing of units involved, victory conditions, and so on. A variety of scenarios is often desirable, but remember that you only play one at a time, so there's no need to read them all before your first game. Similarly, another third of the book might well be historical background of the battle, designer's notes about how the game came to be and why it looks the way it does, and advice on tactics and strategy—much of it interesting, no doubt, but not necessary the first time around.

With a little practice, a first examination of the game map can give you a good idea of the situation and a feel for the importance of terrain. If you're a novice, though, you won't know what to look for; the game's Terrain Effects Chart will be easier to understand.

Your best bet, however, is probably elsewhere. Somewhere at the beginning of the rules, possibly after a lengthy section of historical notes, there will be a brief description of the game situation and, shortly afterward, a general outline of the turn sequence. These two items are the key to the game.

If the general aims of each player are not clear from the basic game situation, you may want to skip down toward the end of the rules section proper and study the victory conditions. With those in mind, you can go back and read the rules through from the beginning. The first time around, however, ignore all sections dealing with optical rules or an "advanced game." Generally, the rules will be grouped by category; for example, Movement Rules, Combat Rules, and so on. Ideally, these groups are arranged roughly in the sequence in which you would encounter them in the course of a normal turn. For instance, movement comes before combat, so the movement rules ought to precede rules dealing with combat. If the rules are lengthy, there may be an outline or table of contents to help you further.

Unfortunately, many rule books are not clearly written. Some contain loopholes, omissions, and even contradictions. This is all the more reason to take it gradually at first. If you have never played a wargame before, your first few games should have a low complexity rating. With a bit of practice, you can work up to games of moderate complexity, but anything more than that requires experience and time.

The place to begin is in the next chapter with *Kassala*, an original wargame carefully designed to be a clear and enjoyable introduction to the world of the deadly die and the cardboard counter.

CHAPTER 4

Kassala: An Introductory Wargame

IT'S ALL very well to read about wargames, to have the common terms and procedures defined and explained, but there's no substitute for seeing and playing one. Hence, *Kassala*.

Kassala is a simulation—a re-creation of an actual, if obscure, sixteenth-century battle that took place between Moslem and Christian soldiers in northeastern Africa. The playing pieces represent the forces—infantry, cavalry, and artillery—believed to have been there. The wargame board is a modified map of the site of the battle.

This game is a short, enjoyable, and challenging contest for two people. The "Moslem" player must attack and try to take the towns of Kassala and Udaka before the game's ten-turn time limit runs out. The "Christian" player must try to prevent this. Although the luck of the die can be a factor, victory will tend to go to the person who outplans and outmaneuvers his opponent.

Finally, *Kassala* is an introduction to contemporary wargames. Wargaming fanatics who disdain anything less elaborate than *War in the East* may not recognize *Kassala* as belonging to the same genre. It isn't, admittedly, a highly refined version of the German invasion of Russia in World War II or Napoleon's defeat at Waterloo. But you'll find all the basic wargaming elements discussed in the previous chapter: a mapboard divided into hexagonal spaces, cardboard playing pieces with printed combat and movement factors, different unit types, zones of control, terrain effects, a Combat Results Table for resolving battles, and so on. To be sure, *Kassala* has fewer pieces—and fewer kinds of pieces—than chess and takes less time to play—twenty minutes or so, with a bit of practice—and the rules have been written specifically to make them easy to understand, none of which is true of the games discussed in the rest of this book. And, once you have mastered the basic game, you can use some of the optional rules that have been provided to add further complexity to *Kassala*. After that, you'll be ready for *War at Sea, Stalingrad, Richthofen's War,* Metagaming's MicroGames, Series 120 games from Game Designers' Workshop, and Simulations Publications' folio and capsule games. Moving up to really complex games like *Squad Leader, Terrible*

Swift Sword, or—if you must—even *Drang Nach Osten!* just takes practice and a little effort.

HISTORICAL BACKGROUND TO THE GAME

In 1527 the first Portuguese embassy to the Coptic Christian kingdom of Ethiopia concluded its visit and sailed for home. Not long thereafter, Imam Ahmed ibn Ibrahim el Ghazi, called Ahmed Gran (the left-handed), led a mixed force of fanatic Moslem warriors on a religious jihad into Ethiopia. Due primarily to the presence within their ranks of two hundred Turkish infantrymen armed with matchlocks, the invading Moslems had little trouble crushing native resistance. Ethiopian Emperor Lebna Dengel scarcely had time to send a plea for aid to Portugal before fleeing to the highlands in the northwestern portion of the kingdom.

For more than a decade the Moslem occupation continued. As the Moslems spread from their homelands on the eastern coast of Africa throughout the kingdom, the Ethiopians were pressed farther back into the highlands. During this period Lebna Dengel died and, in 1540, was succeeded on the throne by his son Claudius, who became Emperor Galawdewos.

In 1541, Portuguese soldiers, armed with muskets and led by Christopher da Gama (son of the explorer Vasco da Gama), slipped into the port of Massawa. Marching to aid the emperor, the Portuguese survived some minor skirmishes against small Moslem bands but were finally attacked en masse by the bulk of the Imam's forces. The Moslem army killed nearly half the Europeans, including Da Gama, but suffered heavy losses. While the Moslems regrouped, the remaining Portuguese hurried to join the Christian forces.

At the walled desert town of Kassala on the river Gash, in what is now Sudan, the forces of Emperor Galawdewos are believed to have made their climactic stand. Reinforced by the veteran Portuguese musketeers and armed from a secret cache of weapons hidden earlier, the Christians prepared their defenses. With reinforcements of their own in the form of troops from the Moslem Ottoman Empire, the army of Ahmed Gran camped less than half a mile away.

At dawn, the Moslem banners were raised; their cannon spouted thunder and lightning, and the assault began. When the clouds of battle finally cleared, the battered Christians had weathered the storm. Most of the Imam's infantry lay dead or dying, and the surviving Moslem forces withdrew to the safety of Khartoum, about two hundred fifty miles to the west.

For two years following the battle, the emperor recruited troops. Then, with a force of eight thousand infantry and five hundred cavalry, he attacked the Moslems at Waina Rega. After sixteen years in Ethiopia, Ahmed Gran fell to a Christian musket ball. Although Emperor Galawdewos died two years later in a minor engagement, he had turned the Moslem tide for good at Kassala.

Kassala: Rules of Play

I. COMPONENTS

A. The mapboard (Fig. 4–1) depicts the area surrounding the town of Kassala, the

KASSALA

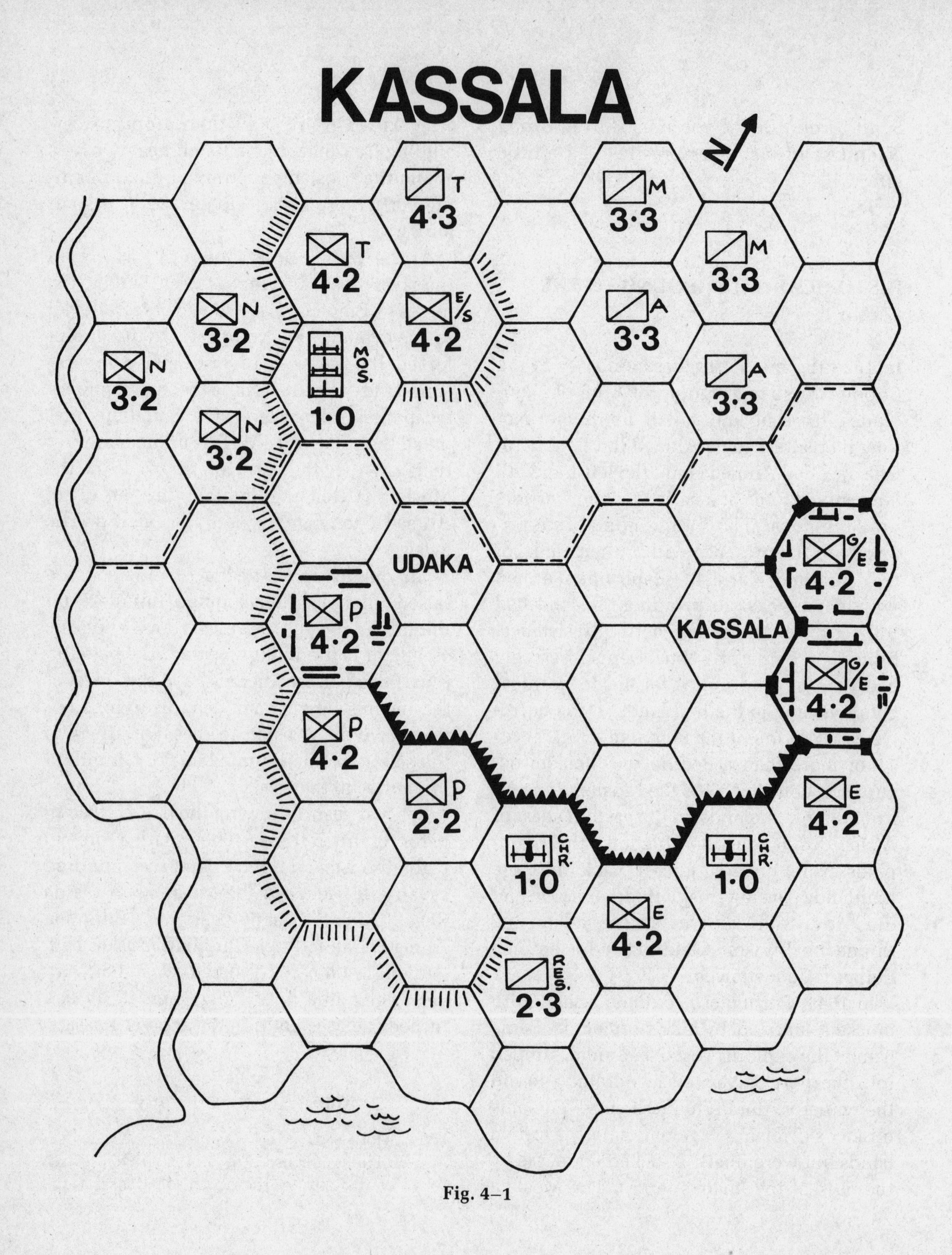

Fig. 4–1

scene of a crucial battle in 1541 between the Christian Ethiopians of Emperor Galawdewos and the invading Moslem army of Imam Ahmed Gram. The terrain has been altered to conform to the strictures of map size. The mapboard has been divided into hexagonal spaces—hexes—to regulate movement and combat. Each hex represents approximately one hundred yards from side to side, and each turn in the game corresponds very roughly to thirty minutes of real time.

1. There are two types of special hexes that affect play.
   a. Kassala Hexes
   b. Udaka Hex
2. There are two types of special hex-sides that affect play.
   a. Wadi Hex-sides
   b. Trench/Abatis Hex-sides
3. The effects of these terrain features are fully explained in the rules for Movement (IID) and Combat (IIE), and are summarized in the Terrain Effects Key.

B. The playing pieces or units are cardboard tiles, often called "counters," which represent the groups of soldiers involved in the conflict. There are two sets of pieces, one for each player. The pieces are listed with their historical equivalents. The number to the left of each unit indicates the total number of each piece. The composition of forces is included for information purposes only; it is not involved in the play of the game.
   1. The Moslem player's units consist of:

(Fig. 4–2) Approximately 300 Marmaluke warriors armed primarily with lances and bows.

(Fig. 4–3) Approximately 200 Arab cavalry armed with assorted weapons.

(Fig. 4–4) Approximately 200 Turkish cavalry with chain-mail armor and assorted weapons.

(Fig. 4–5) Approximately 350 Nubian infantry armed with spears, swords, and bows.

(Fig. 4–6) Approximately 200 Turkish infantry with firearms.

(Fig. 4–7) Approximately 200 Egyptian/Sudanese infantry with firearms.

(Fig. 4–8) One cannon.

2. The Christian player's units consist of:

(Fig. 4–9) Approximately 200 Galla-Ethiopian warriors armed with muskets.

(Fig. 4–10) Approximately 400 Ethopian warriors armed with spears, swords, and bows.

(Fig. 4–11) Approximately 200 Portuguese soldiers armed with muskets.

(Fig. 4–12) Approximately 100 black Portuguese slaves armed with muskets.

(Fig. 4–13) Approximately 150 native light cavalry armed with assorted weapons.

(Fig. 4–14) One cannon.

3. Among both sets of pieces are three types of units. They are distinguished by these symbols in their upper left-hand corner:

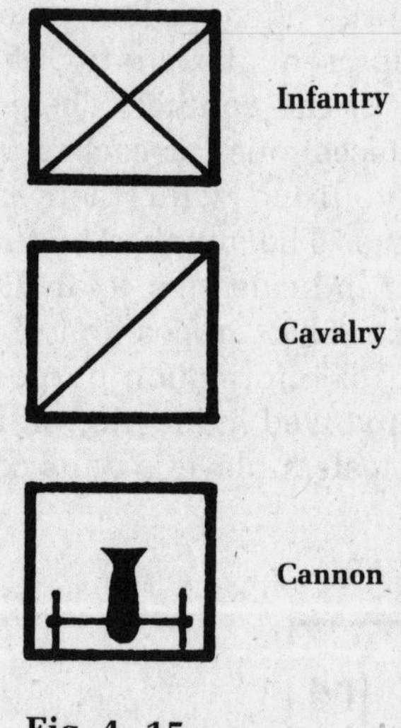

Fig. 4–15

4. The abbreviation in the upper right-hand corner of each piece is used only for historical identification.

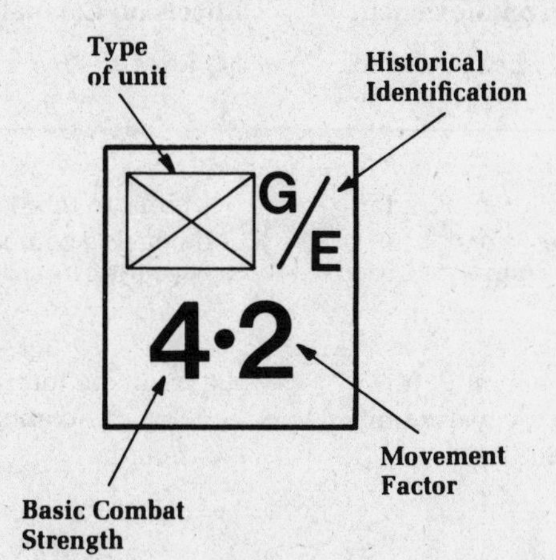

Anatomy of a playing unit or counter.

Fig. 4–16

5. The number in the lower left-hand corner of each piece indicates its *basic combat strength* or the combat factor—the relative strength of a unit when it is attacking or defending against other units. This is very important to the play of the game. In *Kassala*, these combat strengths range from a low of 1 (for cannon) to a high of 4.
6. In the lower right-hand corner of each piece is its *movement factor* or the number of *movement points* it has. Since, generally, it takes one movement point to go from one hex to an adjoining hex, the movement factor is also the basic number of hexes a unit may move in one game turn (every turn). Terrain effects, however, may affect a unit's move, reducing it below this maximum. This, too, is very important. In *Kassala*, the movement factor is 0 for cannon (which can't move), 2 for infantry, and 3 for cavalry units.

C. A die is required to play *Kassala* and must, therefore, be bought or borrowed from another game. Note that the die is used for the resolution of combat only and has no connection whatsoever with the movement of units.

D. The *Combat Results Table* (Fig. 4–17) is used to resolve combat whenever opposing units are next to each other.

Fig. 4–17

Combat Results Table

SIMPLIFIED COMBAT ODDS

DIE ROLL	1–4 or less	1–3	1–2	1–1	2–1	3–1	4–1	5–1 or more
1	C	EX	EX	C	DE	DE	DE	DE
2	C	C	EX	EX	C	DE	DE	DE
3	AE	C	C	EX	EX	C	DE	DE
4	AE	AE	C	C	EX	EX	C	DE
5	AE	AE	AE	C	C	EX	EX	C
6	AE	AE	AE	AE	AE	C	EX	EX

EXPLANATION OF COMBAT RESULTS:

DE = Defender Eliminated. All defending units are removed from the mapboard. Any one attacking unit may advance into each hex vacated by an eliminated defending unit. (Important: These and all other results apply *only* to those units involved in the particular combat being resolved.

AE = Attacker Eliminated. All units participating in this attack are removed from the mapboard.

EX = Exchange. All units of the smaller force (attacking *or* defending) involved in this combat are removed from the board. The owner of the larger force must then remove a number of units whose combined combat strength is *at least* equal to the total combat strength of the smaller force. The larger force's player may choose which units he wishes to eliminate, but those eliminated must be among those—and *only* among those—that were involved in the combat. These strengths are calculated at the "face value" of the units and, for this purpose, are not modified by terrain additions or subtractions. Note that an "Exchange" may result in the elimination of *both* forces. Any surviving *attacking* units (only) may occupy any hexes vacated by eliminated *defending* units (this, of course, does not apply if all the attacking units are removed and some defending units remain).

C = Contact. There is no effect to either the attacking or the defending units unless the defending hex is occupied solely by cannon units. In this case, treat the "Contact" result as "Defender Eliminated" in regard to the cannon. In all other cases, all involved units remain in place and must renew the battle in the combat portion of the other player's turn. Additional units may join the units in Contact, but no unit in Contact may move away.

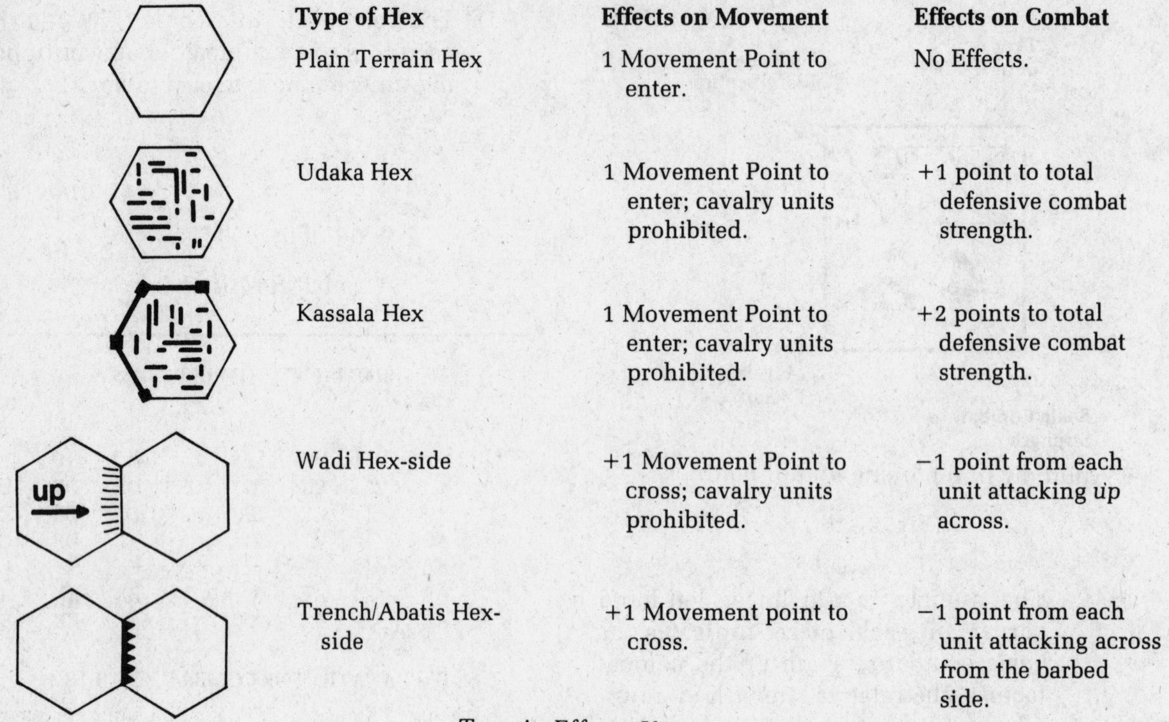

Type of Hex	Effects on Movement	Effects on Combat
Plain Terrain Hex	1 Movement Point to enter.	No Effects.
Udaka Hex	1 Movement Point to enter; cavalry units prohibited.	+1 point to total defensive combat strength.
Kassala Hex	1 Movement Point to enter; cavalry units prohibited.	+2 points to total defensive combat strength.
Wadi Hex-side	+1 Movement Point to cross; cavalry units prohibited.	−1 point from each unit attacking *up* across.
Trench/Abatis Hex-side	+1 Movement point to cross.	−1 point from each unit attacking across from the barbed side.

Terrain Effects Key

Fig. 4–18

E. The *Terrain Effects Key* (Fig. 4–18) summarizes the effects, on both the movement of units and combat, of the different hexes and hex-sides on the mapboard.

II. How to Play the Game

A. Setting Up the Game

1. Put the mapboard on the table or other flat, solid surface, and place each playing piece on the hex bearing that unit's symbol and historical designation. Note that the three Moslem cannon are stacked in the same hex.

B. Sequence of Play

1. The player controlling the Moslem units moves any or all of his units within the limits of their movement factors and the rules for movement.
2. When he has moved all the units he wants to move that turn, the Moslem player then makes any attacks required by the rules for combat; that is, his units must attack any Christian units to which they are adjacent.
3. When the Moslem player has completed all attacks for the turn, the Christian player moves any or all of his units within the limits of their movement factors and the rules for movement.
4. The Christian player then makes any attacks required by the rules for combat.
5. The players record the passage of one game turn.
6. This sequence of play is repeated ex*actly* for ten game turns; at the end of the tenth turn, the outcome is determined.

C. Zones of Control

1. In addition to occupying the hex on which it rests, a unit in *Kassala* also exerts an influence on every adjacent

hex. This is the unit's *zone of control*.

2. This influence affects the movement of enemy units and affects combat for all units. These effects are fully explained within the rules for Movement and Combat.

D. Movement

1. During any turn, a player can move as many or as few of his units as he wishes.
2. A player doesn't have to move a unit as far as its movement factor would allow. For example, a cavalry unit with a movement factor of 3 might, on any particular turn, move 1, 2, or 3 hexes— or not at all.
3. Unused movement points cannot be saved for a later turn or transferred to another unit; not moving one unit does not allow you to move the one next to it twice as far as normal.
4. A unit can be moved every turn; its movement factor is the same throughout the game, no matter how much or how little the unit is moved.
5. Basically, it takes one point of a unit's movement factor to move it from one hex to an adjacent one, but note the exception in the following rule.
6. Crossing a wadi or trench/abatis hex-side costs an additional movement point. Thus, it takes *two* movement points to go from one hex to another across a wadi/trench/abatis hex-side. (See Examples of Play.)
7. No unit may enter a hex unless that unit has enough (unused) movement points to do so. For example, an infantry unit with 2 movement points cannot move one hex and then cross a wadi hex-side into a second hex; that would require 3 movement points, not 2. Instead, it must stop after one hex and, during the player's *next* turn, use its full movement factor to cross the wadi hex-side to the hex beyond.
8. In the movement portion of any single game turn, no unit may be moved a number of hexes greater than its movement factor.
9. Units may, at no extra cost in movement points, move *through* hexes that are occupied by friendly units, but no two units may occupy the same hex at the end of their movements. In other words, "stacking," which is allowed to some extent in other games, is not allowed. Note: The stack of three Moslem cannon units is the *only* exception to this rule. Units may never enter hexes occupied by enemy units.
10. A unit must cease movement as soon as it moves adjacent to an enemy unit and *may not move again*—even in later turns—until all enemy units adjacent to it are destroyed in combat.
11. Units beginning movement adjacent to an enemy unit may not move during that game turn. The only exception is detailed in the rules for combat (E13).
12. Cavalry units may never be moved into or through any Kassala or Udaka hex at any point in the game.
13. Cavalry units may never be moved across any wadi hex-side at any point in the game. They may, however, cross trench/abatis hex-sides in the same way that infantry units do —at the cost of an extra movement point.
14. Cannon units may never be moved at any point in the game.
15. Units may never be moved off the mapboard or onto any portion of the board that is not a hex (e.g., the river).
16. Once a player has moved a unit and removed his hand from it, he cannot alter that unit's move without the consent of his opponent.
17. Movement as a result of combat is a

special case and is described in combat (E13).

E. Combat

Combat is required whenever one or more Christian units are adjacent to one or more Moslem units. The player whose turn it is (normally, the one who just moved) is the attacker; the other player is the defender (for this portion of the turn only; in *his* turn, he will become the attacker). Combat is resolved by comparing the total combat strength of the attacking units (in each particular battle) to the total combat strength of the units defending against that attack. This comparison is simplified—rounded off in favor of the defender—and expressed as a ratio whose smaller number is 1 (one). This ratio is then used to determine which column of the Combat Results Table (CRT) to use for the battle.

Example: A unit with a combat strength of 4 attacks an enemy unit with a combat strength of 2. Simplifying the comparison of 4 to 2 yields a ratio (or odds) of 2 to 1. The attack is resolved on the 2–1 column of the Combat Results Table.

Example: Units with a total combat strength of 5 attack an enemy unit with a combat strength of 3. Since all ratios are rounded off in favor of the defender, this attack is resolved on the 1–1 column of the Combat Results Table, although the actual ratio is closer to 2–1.

After determining which column of the Combat Results Table to consult, the player attacking rolls a die and matches the number shown, with the correct column of the CRT. The result—AE, DE, EX, or C—is applied immediately to the battling units. (The results are explained in the CRT.)

1. A unit may only attack adjacent enemy units.
2. Each adjacent enemy unit may be attacked only once each turn, although in the other player's turn the attacked unit may—and in some cases *must*—make its own attack.
3. All enemy units adjacent to friendly units at the beginning of the combat portion of your turn *must* be attacked that turn. All friendly units that begin combat adjacent to an enemy unit *must* participate in an attack that turn.
4. Each unit may attack only once each turn (the converse of Combat Rule No. 2).
5. A unit may, however, attack more than one adjacent enemy unit at the same time (i.e., in the same attack).
6. A unit's combat strength is an integral whole and may never be split and applied to more than one combat in the same game turn. (This is analogous to Movement Rule No. 3.)
7. A player may choose which of his units will attack each enemy as long as the preceding rules are followed.
8. When defending (only), the *total* combat strength of *all* units in *each* Kassala hex is increased by 2 points (e.g., from 4 to 6). For Udaka, the defensive bonus is *one* point. (Note: If the optional cannon rules are used, this bonus applies to the total strength of a stack, not to each unit in a stack.)
9. Each unit attacking *up* across a wadi hex-side loses one point from its combat strength for that attack. (Important: Note that this is a "one-way" effect and does not apply to units attacking from an unmarked hex *down* across a wadi hex-side. See the Terrain Effects Key.) This rule does not apply to cannon.
10. Each unit attacking across a trench/abatis hex-side *from the barbed side* (only) loses 1 point from its combat strength for that attack. (Rules No. 9 and 10 apply if a unit is making a combined attack against two or more units, some of which lie across

trench/abatis hex-sides.) This rule does not apply to cannon.

11. Cavalry units may participate in attacks on units in Kassala or Udaka hexes and across wadi hex-sides but may not enter such hexes or cross such hex-sides at any point in the game.
12. The attacking player may resolve combats in any order desired, but the result of each combat must be applied immediately.
13. If, as a result of combat (i.e., DE or EX), a defending unit is eliminated, any (one) surviving attacker may immediately be moved into the hex vacated by the eliminated defender. This applies only to a unit which actually participated in the attack that eliminated the defender. Note that this move is over and above any normal movement allowed by a unit's movement factor. Note too, however, that cannon cannot move, and that cavalry still cannot enter Udaka or Kassala or cross wadi hex-sides.
14. For further clarification, see the Combat Results Table and the Examples of Play.

III. Winning the Game

A. The outcome is determined at the end of ten complete turns of play.

B. There are three possible results of a game of Kassala:

1. The Moslem player wins if there are no Christian units in any Kassala or Udaka hex at the end of the game.
2. The Christian player wins if he has units *in both* Udaka and Kassala at the end of the game.
3. The game is a draw if *either* Kassala or Udaka (but not both) is vacant of Christian units at the end of the game.

IV. Optional Rules

If, after learning the basic version of *Kassala* described above, you wish to play a more challenging game, you may add any or all of the following optional rules. (All the basic rules still apply unless they are specifically contradicted.)

A. Cannon

1. Cannon units are special units and function differently from other units.
2. Unlike other units, cannon may be stacked in a hex, either with other cannon or a different kind of unit or both. As many as three cannon and one other unit (cavalry or infantry) may occupy the same hex. (They would have to be all Moslem or all Christian units.)
3. Each cannon in a hex contributes one strength point to the defense of that hex.
4. Any "Contact" result against a hex defended solely by cannon is treated as though the result had been "Defender Eliminated" (DE).
5. Cannon units may attack enemy units up to two hexes away.
6. Cannon may only attack enemy units that are adjacent to friendly units. (Note this restriction well.)
7. Cannon can perform three types of attacks, but each unit may only participate in one attack during any game turn.

   a. *Combined Attacks* are simply those in which one or more cannon combine with one or more other units to attack one or more enemy units. Any cannon participating in such combat must be within two hexes of at least one of the units being attacked. Each cannon may add its single point of combat strength to only one attack each turn. Cannon units never suffer any adverse effects from participating in combined attacks against *nonadjacent* enemy units (i.e., they cannot be eliminated as a result of AE or EX, and "Contact" results would not apply to them).

   b. *Bombardment* is a solitary attack by a cannon against an enemy unit adjacent to another friendly unit.

Its purpose is to allow the friendly unit to concentrate its combat strength against other adjacent enemy units. (The cannon is making a "diversionary attack" or, in effect, "soaking off.") Remember, however, that every friendly unit that begins the combat portion of a turn adjacent to an enemy unit must participate in an attack that turn. Bombardment attacks *never have any effect* on the enemy units being attacked. The die is not rolled, and the CRT is not consulted.

c. Cannon may occasionally be forced to make an *Individual Attack* against an *adjacent* enemy unit or units. Results of these attacks are always treated as "Contact" results (but note that this is *not* the same as Rule IV A4), and the die is not rolled. (Note, however, that due to the rule just cited, the next enemy attack *will*, presumably, eliminate the cannon.)

8. No terrain effects reduce, inhibit, interfere with, or eliminate the combat strength of a cannon or any attack made by one. But note that the normal defensive bonus of occupying a Kassala or Udaka hex applies to cannon or to any unit(s) being attacked by cannon.

B. Free Deployment

This allows players to vary the game setup from the basic positions printed on the mapboard.

1. The Christian player places his units in any hexes south of the dotted line on the mapboard.
2. The Moslem player then sets up his units in any hexes north of the same dotted line.

C. Mixed Attacks

Whenever one or more infantry units make a joint attack with one or move cavalry units against one or more enemy units, the attack is resolved on the Combat Results Table *one column higher* than would normally be the case. For example, if the Turkish 4·3 cavalry and the Turkish 4·2 infantry were attacking one of the Christian 4·2 infantry units at 2–1, the attack would be resolved on the 3–1 (*not* the 2–1) column of the CRT. Note that this bonus applies only to *attacking* units and is not dependent on the composition of the defending units (which could be a mixture of infantry and cavalry units, also, without altering the attacking force's bonus).

V. Examples of Play

A. Movement (Fig. 4–19)

The Nubian infantry may move to either hex (a) or (b) at a cost of two movement points (one for the hex entered and one for the type of hex-side crossed). The unit may, alternatively, move to hex (c) via hex (d) at a cost of one movement point for each hex entered.

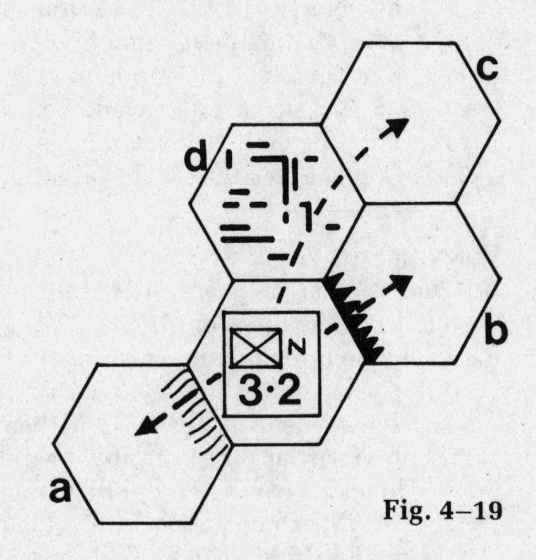

Fig. 4–19

B. Single Attack (Fig. 4–20)

The Turkish infantry attacks the enemy unit in hex (b). The Turkish unit (a) loses one strength point for attacking up across a wadi hex-side. The Portuguese unit gains one strength point for defending in the Udaka hex. The adjusted attack is thus 3–3 and is resolved at 1–1.

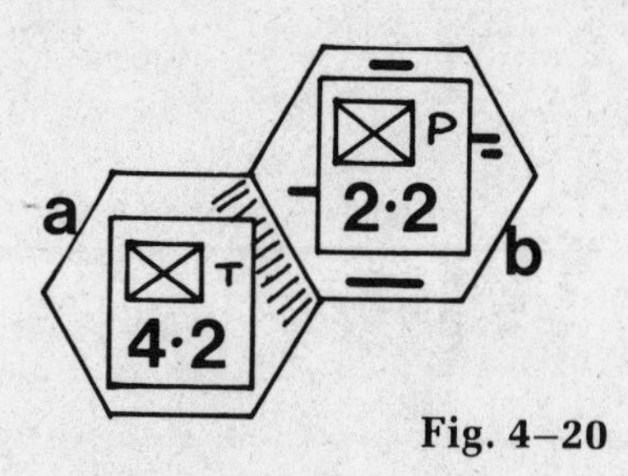

Fig. 4–20

C. Multiple Attack (Fig. 4–21)
The Egyptian infantry (a), the Turkish infantry (b), the Turkish cavalry (c), and the Arab cavalry (d) attack the Ethiopian infantry (f) and the cavalry reserve (e). The attack must be executed in one of the following ways:

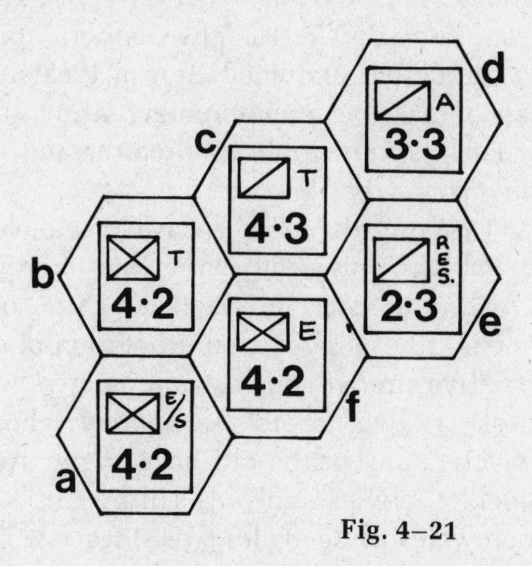

Fig. 4–21

1. The Turkish and Arab cavalry (c) and (d) may attack the reserve cavalry (e) at 3–1 (7–2) odds, while the remaining Moslem units (a) and (b) attack the Ethiopians (f) at 2–1 (8–4) odds.
2. Alternately, the Arab cavalry (d) may attack alone against the reserve cavalry (e) at 1–1 (3–2) odds, while the remaining Moslem units (a), (b), and (c) attack the Ethiopians (f) at 3–1 (12–4) odds (if the optional rules were being used, this second attack —but not the combat between the cavalry units—would be a *mixed attack*).
3. Note that the four Moslem units could not make one large joint (and mixed) attack against the two Christian units, since the Arab cavalry (d) is not adjacent to the Ethiopian infantry (f), and the two Moslem infantry units (a) and (b) are not adjacent to the Christian reserve cavalry (e). (Combat Rule No. 1.)

D. Bombardment (Fig. 4–22)
If the Moslem cannon (d) bombards the Portuguese infantry (b), the Turkish infantry (c) is free to attack the cavalry reserve (a) at 2–1 (4–2) odds. If the cannon does not bombard the Portuguese infantry, the Turkish infantry will be forced to attack both of the Christian units at 1–2 (4–6) odds.

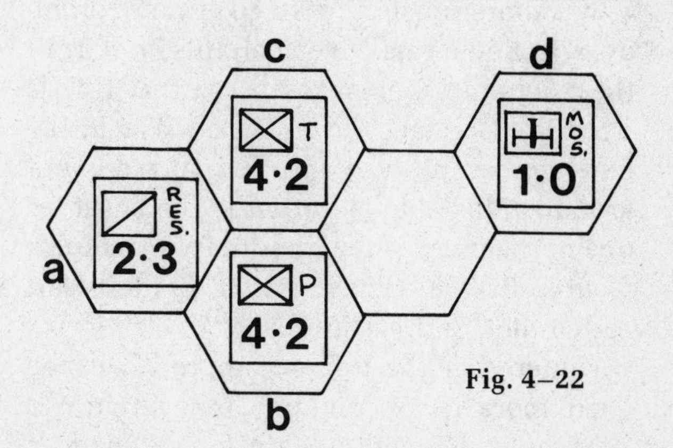

Fig. 4–22

CHAPTER 5

Playing to Win

THERE'S MORE to playing a wargame than learning the rules. Whether you're playing against someone else or just by yourself, you'll have to understand strategy and tactics to do it right.

Strategy can be described as where we're going or what we're going to do; tactics involve how we get there or do what we want. Underlying both terms are certain basic military principles which, unsurprisingly, are relevant to many wargames.

A frontal assault, for example, can be compared to ramming your head into a brick wall. If you happen to be the Incredible Hulk, this approach may work. If not, you'll find it more satisfactory to go around, climb over, or find a gate. If you call this brick wall the Maginot Line, you will see that this is what the Germans did in World War II: the *Luftwaffe* went over, and the panzers went around. This simple notion is the basis for one of the most essential military principles —what British strategist B. H. Liddell Hart called the "indirect approach."

Underlying the indirect approach is the even more basic principle of bringing to bear your strengths against the enemy's weak points. The rear or flank attack, an obvious example of the indirect approach, illustrates this point. The spears and shields of pregunpowder armies faced the front; men could not see—much less protect themselves from—someone attacking from the rear. Hence, the importance of "facing" in a game like *Alexander the Great* or *The Battle of Raphia* and the bonus given thieves in *Dungeons & Dragons* for striking from behind. The fact that modern combat is normally conducted from a greater distance makes a contemporary army somewhat less vulnerable to a rear assault—but only somewhat.

"Crossing the T" is a naval maneuver in which the massed fire of a line of ships is brought to bear on the lead ships of an enemy line. This is a confrontation of comparative strength against comparative weakness on two levels. As anyone who has watched any of the old Errol Flynn swashbuckler movies knows, a sailing ship's cannon were arranged along its sides; few if any guns were pointed toward the bow or stern. The main batteries of a twentieth-century battleship could swivel from side to side, but the aft guns could not fire forward and the guns forward could not fire aft. Thus, a line of ships crossing ahead of another line of ships could fire broadsides against vessels that could not bring the full strength of their own guns to bear on the attackers. Also, the basic geometry of the situation allowed more of the crossing ships to exchange fire with fewer elements of the "crossed" line; those in the rear—perhaps the vast majority—could do nothing but watch the ships in front get blown apart.

The ships of antiquity fought in a different way, ramming their heavy reinforced prows into the vulnerable sides of enemy galleys. The principle, however, was the same: strength against weakness.

This maxim can explain the traditional reluctance of a commander to split his forces and risk what is termed "defeat in detail." By applying its full strength in turn against the lesser strengths of the separate units of a larger army, an undivided force can achieve a victory that would have been difficult or impossible in a single great battle. It is sometimes suggested that the true comparison of the strengths of two forces that are similar in composition but different in size is the ratio of the square of their numbers; for example, a two-to-one majority may yield a practical advantage of four to one. This is certainly true in wargames. Attacking a force of several units as a group on a one-to-one basis is generally unproductive and often suicidal. But attacking each unit one at a time at a three- or four-to-one advantage will, after several turns, give you a complete and inexpensive victory.

A cautionary note about economy of force is necessary: there's not much point in using a machine gun to kill mosquitoes. If you have several objectives, you may have to split your forces, and you rarely have the luxury of mustering overwhelming force against each of the enemy positions. If a three-to-one attack will do the job, don't bother trying to increase the advantage to six to one; it can get you in trouble somewhere else.

SIGNPOSTS AND SPEED LIMITS

In a game, the conditions for victory are basic determinants of strategy; the basic determinant of tactics is the Combat Results Table.

In planning your strategy, the first thing to do is to examine carefully what is necessary for a game victory. Even experienced gamers often lose sight of this. A Moslem player in *Kassala* may conduct a tactically brilliant campaign, defeating every Christian unit he attacks. However, if after ten turns there are Christian units left in the towns of Udaka and Kassala—even if they're the only two remaining Christian units—the Moslem player has lost the game even if he hasn't lost a single piece. "Winning the battle and losing the war" is not an empty phrase. Despite objections from careless players and disgruntled losers, there is abundant historical justification for this seeming anomaly. There are a few cases in which a historically absurd victory may win the game, but remember that a game is what you're playing: the actual wars of the past are beyond your reach. If the victory conditions displease you, talk your opponent into changing them before you start the game—or play alone.

The tactics of combat will largely depend on the nature of the Combat Results Table. It is hardly feasible to examine each of the CRTs currently in use, but many have certain key features in common, and in any case the lessons derived from a few examples can be applied to those with different numbers.

Examine the Combat Results Table from *Kassala* (Fig. 5–1). Note first the obvious facts that an attack at 1–4 odds will probably accomplish nothing except the defeat of the attacker, while an attack at 5–1 will almost certainly eliminate the defender. At other odds, this table, like all CRTs, shows a fairly straightforward progression from one extreme to the other: the better the odds, the greater the attacker's chances for victory.

Fig. 5–1

Combat Results Table

SIMPLIFIED COMBAT ODDS

DIE ROLL	1–4 or less	1–3	1–2	1–1	2–1	3–1	4–1	5–1 or more
1	C	EX	EX	C	DE	DE	DE	DE
2	C	C	EX	EX	C	DE	DE	DE
3	AE	C	C	EX	EX	C	DE	DE
4	AE	AE	C	C	EX	EX	C	DE
5	AE	AE	AE	C	C	EX	EX	C
6	AE	AE	AE	AE	AE	C	EX	EX

EXPLANATION OF COMBAT RESULTS:

DE = Defender Eliminated. All defending units are removed from the mapboard. Any one attacking unit may advance into each hex vacated by an eliminated defending unit. (Important: These and all other results apply *only* to those units involved in the particular combat being resolved.

AE = Attacker Eliminated. All units participating in this attack are removed from the mapboard.

EX = Exchange. All units of the smaller force (attacking or defending) involved in this combat are removed from the board. The owner of the larger force must then remove a number of units whose combined combat strength is *at least* equal to the total combat strength of the smaller force. The larger force's player may choose which units he wishes to eliminate, but those eliminated must be among those—and *only* among those—that were involved in the combat. These strengths are calculated at the "face value" of the units and, for this purpose, are not modified by terrain additions or subtractions. Note that an "Exchange" may result in the elimination of *both* forces. Any surviving *attacking* units (only) may occupy any hexes vacated by eliminated *defending* units (this, of course, does not apply if all the attacking units are removed and some defending units remain).

C = Contact. There is no effect to either the attacking or the defending units unless the defending hex is occupied solely by cannon units. In this case, treat the "Contact" result as "Defender Eliminated" in regard to the cannon. In all other cases, all involved units remain in place and must renew the battle in the combat portion of the other player's turn. Additional units may join the units in Contact, but no unit in Contact may move away.

Some differences between one column and the next can be more significant than others, however. Notice that at 2–1 odds there is still a chance that the attacker will lose his entire force; since by definition he has more at stake than the defender and may, due to rounding, have almost three times the defender's troops, this is a very risky attack and one that should be avoided. An attack at 3–1 odds is much better because there is twice the chance that the defender will be eliminated outright. More important, the attacker cannot lose; at worst, he may have to sacrifice a third of his units. Most CRTs—and all of the CRTs in Avalon Hill's older games—show a similar breakpoint at 3–1 odds.

Since a "C" result at high odds only postpones the inevitable for half a turn, a 4–1 attack is not significantly better than one at 3–1. Since there is no "stacking," which allows a greater concentration of force and, therefore, the possibility of better odds, in *Kassala,* attacks at 4–1 or 5–1 odds are nearly impossible anyway. Even in other games, as we have noted, they are likely to be impractical.

A 1–1 attack is better for the defender, but not significantly worse for the attacker than odds of 2–1; consequently, a combination of a 3–1 and 1–1 attack is much better for the attacker than a pair of risky 2–1 attacks. At times, to get good odds for one or more attacks, it may be necessary for one unit to fulfill the legal combat requirements by making a diversionary attack at poor odds against other enemy units. This maneuver is sometimes called "soaking off." In *Kassala,* artillery units are ideal for this role, because if they attack from a distance of two hexes, they cannot be affected by a negative outcome on the CRT.

Other conclusions can be drawn from the nature of the combat system. One application of the principle of economy of force is that it is unnecessary and dangerous to have, for instance, a 4·2 infantry unit and a 3·3 cavalry unit make a combined attack on an enemy 4·2 infantry unit, assuming no terrain modifications are required. The combination can attack at no better odds (1–1) than the infantry unit could manage on its own; the cavalry unit is being risked for no purpose. Generally, you should use only as many combat factors' worth of units as you need to get the odds you want.

There are, however, two occasional exceptions to this rule. One is if the position of the defending unit is important. In *Kassala*, for instance, if the defender is in Udaka, it may be desirable to risk an extra unit so that, in the case of an "Exchange" result on the CRT, the survivor can take the defender's place and occupy the town. An example of the second exception would be a 4·2 unit attacking a 3·2 unit at 1–1 (4–3 rounded) odds. The most likely result, "C," would effectively force the 3·2 defender to counterattack in its turn. However, this 3–4 attack would be rounded to the more dangerous (for the ex-defender) odds of 1–2. Since units lose their defensive terrain bonuses when they attack, the actual odds shift in some cases may be even greater. In *Kassala*, the defensive positions of the Christians are so strong that the attacking Moslems must use this tactic occasionally.

Many other CRTs contain some form of retreat results; for example, "D back 2," which means the defender must retreat two hexes. Some, like the CRT from *Blue & Gray* QuadriGame (Fig. 5–2), are almost nothing but retreats. Unsurprisingly, this will affect proper tactics.

Despite the great differences from the *Kassala* Combat Results Table, this example shows the characteristic advantage of a 3–1 attack (there are no "Attacker Retreats" results). A more important point is related to a rule not directly shown. A unit that is surrounded—either by enemy units or their zones of control—cannot retreat and is elim-

Fig. 5–2

Combat Results Table from Blue & Gray *QuadriGame.*

COMBAT RESULTS TABLE

Die Roll	Probability Ratios (Odds) Attacker's Strength to Defender's Strength										Die Roll
	1-5	1-4	1-3	1-2	1-1	2-1	3-1	4-1	5-1	6-1	
1	Ar	Ar	Dr	Dr	Dr	Dr	De	De	De	De	1
2	Ar	Ar	Ar	Dr	Dr	Dr	Dr	Dr	De	De	2
3	Ar	Ar	Ar	Ar	Dr	Dr	Dr	Dr	Dr	De	3
4	Ae	Ar	Ar	Ar	Ar	Dr	Dr	Dr	Dr	Dr	4
5	Ae	Ae	Ar	Ar	Ar	Ar	Dr	Ex	Ex	Dr	5
6	Ae	Ae	Ae	Ae	Ar	Ar	Ex	Ex	Ex	Ex	6

Attacks executed at greater than 6-1 are treated as 6-1; attacks executed at worse than 1-5 are treated as 1-5.

EXPLANATION OF COMBAT RESULTS

Ae = **Attacker Eliminated.** All Attacking units are eliminated (remove from the map).

De = **Defender Eliminated.** All Defending units are eliminated.

Ex = **Exchange.** All Defending units are eliminated. The Attacking Player must eliminate Attacking units whose total, printed (face value) Combat Strength **at least** equals the total printed Combat Strengths of the eliminated Defending units. Only units which participated in a particular attack may be so eliminated.

Ar = **Attacker Retreats.** All Attacking units must retreat one hex (see 7.7).

Dr = **Defender Retreats.** All Defending units must retreat one hex.

CREDIT: Courtesy of Simulations Publications, Inc.

inated instead. With such a "bloodless" CRT, this is virtually the only way to eliminate units. Under a normal 3–1 attack, the most likely result is a "Dr"—the defending unit retreats one hex but remains intact. If the same 3–1 attack is made when the defending unit is surrounded by attacking units (even those involved in another battle) and/or their zones of control, the defender must be eliminated. And there is only a one-sixth chance that the victory will cost the attacker any units.

Obviously, therefore, with such a Combat Results Table it is very important to surround the defender. It's even more important than improving the odds. This isn't as difficult as it may seem and certainly does not require six attacking units. If the rules prohibit retreats through enemy zones of control, only two attackers—on opposite sides of the defender—are required.

In Fig. 5–3, all the hexes marked "1" are in the zone of control of the first attacking unit (A1); the hexes marked "2" comprise the zone of control of the second attacker (A2). The hex occupied by the defender (D) is, of course, within the zone of control of both attackers. The important point is that every hex adjacent to the defender is blocked by the attacking units or their zones of control. A nominal retreat result will instead result in the defender's elimination.

The order in which multiple battles are resolved is often important, particularly when it determines whether certain defending units are surrounded.

In Fig. 5–4, attacker A3 may combine with A1 to attack D1 or with A2 to attack D2. Since the position is symmetrical, let's assume that A3 combines with A2 to make a 2–1 attack on D2. The exact odds and combat factors are less important than the principle involved; it does no harm to consider all five units to be of equivalent strength. The presence of A3 means that, despite the difference in odds in the two attacks, both defending units are surrounded. Now, regardless of the outcome of the A1 attack on D1, D2 will still be surrounded for the combined attack from A2 and A3. On the other hand, if A2 and A3 were forced to retreat or, worse, eliminated—as might be possible on another CRT—the defending unit D1 (depending on stacking restrictions, length of retreat, and so on), would no longer be surrounded. Consequently, it is crucial to resolve the A1 attack first and then the combined A2–A3 battle.

In other situations, an initially protected unit—part of a line of defensive units—

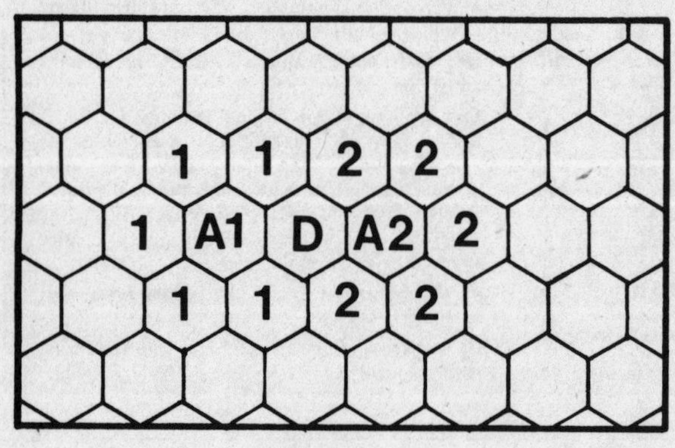

Fig. 5–3

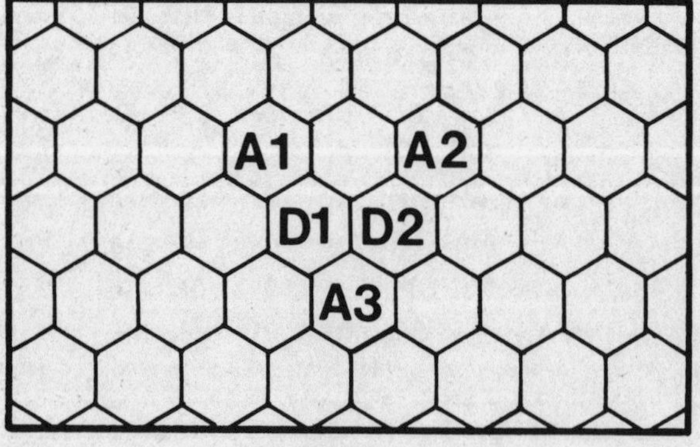

Fig. 5–4

may, in the course of a multiple battle, become surrounded by victorious advancing units, which are often permitted to occupy the hex abandoned by a retreating or eliminated defender. Generally, these units may not actually participate in a second attack in the turn, but their presence and their zones of control may quite legitimately succeed in cutting off the remaining defender. Again, the order in which the individual battles is resolved (in such a case, from flank to center) must be chosen carefully.

MY KINGDOM FOR A HORSE

The individual capabilities of your units will also affect your course of action. In *Kassala,* the Moslem player cannot use any of his large cavalry force to outflank the Christian units in and around the town of Udaka because cavalry units cannot cross the wadi hex-sides. The Christian reserve cavalry could go around—not over—the long line of wadi hex-sides, but this would rarely be desirable. Similarly, since only infantry units can occupy town hexes, it will be very difficult for the Moslem player to push the Christians out of Udaka and Kassala—and keep them out—if his only surviving units are cavalry. If, as might be the case in another game, the Moslems were required to occupy the towns, it would obviously be not just difficult but impossible for the Moslem player to win without preserving some infantry units for that purpose.

Some units in many games are stronger offensively than defensively, or vice versa, and should be used appropriately. The most numerous German units in *D-Day* are 1·2·2 (attack factor 1, defense factor 2, movement factor 2) "static infantry" divisions. They are the weakest units on the board, but despite their limitations, the German player must make good use of them to have any chance of stopping the Allied invasion. They are too weak to leave in the open; they must be combined with one or two others in the same hex and placed in positions that augment their defensive strength: in city and fortress hexes or, at least, strung out behind rivers to slow the Allied advance. If they are feeble defensively, they are hopeless in an attack; counterattacks must be left to the far more powerful panzer divisions. This is another reason for putting static divisions in fortresses, which do not require their "residents" to counterattack adjacent units. And, since the German replacement rate is based only on attack factors, the asymmetrical static infantry units make cheap but effective reinforcements.

In some scenarios of *PanzerBlitz,* the German forces have two units of awesome firepower, the Hummel and the Wespe. Their range and attack factors are far greater than anything else in the game. Defensively, however, they are quite ordinary, and they are much too valuable to risk in a simple exchange of fire with the more numerous and better armored Russian tanks. It is imperative that these vulnerable "superweapons" be placed well back of the front line and on high ground so that they can "see" the length of the board and destroy approaching enemy units before they can get close enough to return fire.

All units must be used to best advantage. If the game you're playing features double-impulse movement for certain pieces, those armored or mechanized units should be placed to take advantage of their second-movement segment. For example, they should be used to exploit any holes in the enemy line created by combat results, by "breaking through" and surrounding the units in the enemy front line. In *Tactics II*, to cite another example, a mountain unit—which, in the mountains, can only be attacked by another mountain unit—blocking the only pass can bring an entire column to an indefinite halt.

If you have a choice of units to begin the game or to bring on as reinforcements later, your selection will depend on where you're going to use the units and for what purpose. It also depends on their particular abilities and their cost in strength points, economic units, or whatever.

In *Diplomacy*, Russia would build a fleet at St. Petersburg (north coast) only if it was opposing England, either directly or through the occupation of the Scandinavian countries. On the other hand, armies built in Moscow and Warsaw might indicate an imminent attack on Germany. If England builds only fleets, the English player must confine himself to the occupation of seas and coastal provinces, an arrangement that might be quite satisfactory in an alliance with a land-based power like Germany or Austria-Hungary.

We have already noted that the German player in *D-Day* can, for defensive purposes, double his effective replacement rate by selecting the offensively weaker 1·2·2 static divisions. There are similar loopholes, advantages, and drawbacks to units in nearly all games. In *Ogre*, the defender generally does better to pick a multitude of weak units rather than fewer strong units, which are equally vulnerable to overruns by the giant Ogre, a cybernetic supertank. In *Alpha Omega*, the reverse is true: large ships are as fast as smaller ones and, by allotting enough energy to the defensive "cloaking" device, are essentially invulnerable to the weapons of a lesser craft. Unit selection in a game like *Imperium* is more complex. Small ships can be overwhelmed by larger ones, but battleships are more expensive to build and, because of their high maintenance costs, difficult to maintain.

DEFENSE: THE FORGOTTEN HALF

Strategy and tactics on defense are also largely determined by a game's victory conditions and terrain. In *D-Day*, the German defense is basically a matter of taking advantage of defensive terrain bonuses—falling back gradually from one river or cluster of city hexes to another. In *Kassala*, the Christians logically will occupy and defend the towns of Udaka and Kassala as long as possible, both because occupation is essential to winning and because, due to their defensive bonuses, the town hexes are strong positions. The other Christian units are shielded in front by the trench/abatis line and protected on the flanks by the towns and the wadi hex-sides. Even if Moslem units manage to pierce either barrier, allowing them to make a combined flank and frontal assault, the terrain bonuses will not be wholly negated, and the lack of room to maneuver will still make it difficult to obtain consistently good odds. Consequently, the Christians should remain behind these barriers and shift only slightly to meet major Moslem thrusts. Counterattacks should be mounted only to maintain the integrity of their defensive positions or to recapture the towns. There is no point whatsoever to venturing onto the open plains to the northwest. The Moslem player, on the other hand, will try to deprive some of the Christian units of their defensive bonuses by penetrating the trench and wadi lines and tying them up with forced counterattacks.

Where terrain is not a consideration, and ranged fire is not typical of the units in the game, there are certain basic defensive doctrines that follow largely from the nature of the hex grid. A great many games involve zones of control, allow victorious units an advance after combat, and have some sort of retreat results in the Combat Results Tables, which makes it advantageous to surround a defending unit.

In these circumstances, the defending player should, if at all possible, arrange his units in a straight line "along the grain" of the hex grid, two hexes apart so that they occupy every other hex. A line perpendicular to any one of the sides of a hex will lie along the grain of the grid or, more accurately, along one of the three grains of the hex grid. This is contrasted to the uneven pattern traced by a line going out of one of the corners of a hex.

Assuming that a defensive line of counters (Fig. 5–5) cannot be outflanked, each unit in it can be attacked by only two enemy units (or stacks of units, if stacking is allowed), such as those in the positions marked A. The hexes directly between the defending units cannot be occupied by attacking units, which are stopped by the defenders' zones of control as soon as they reach position A. Furthermore, even if one unit falls and its position is taken by an attacker, the nearest defenders are still not surrounded. A reserve kept behind this line can drive off or eliminate the invading unit and restore the line.

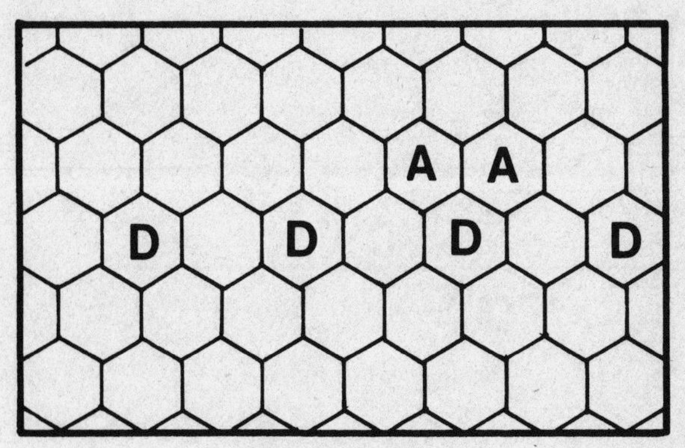

Fig. 5–5

While the small mapboard, the presence of favorable terrain, and the absence of retreat results make it feasible in *Kassala*, a solid line of adjacent units is not usually a practical defense. For one thing, the defending player doesn't normally have enough units available. For another, victorious units advancing after combat into such a line could easily surround some of the remaining units. Furthermore, overlapping zones of control would force the threatened defending unit to counterattack—possibly at unfavorable odds.

On the other hand, a defensive line in which defending units are separated by two or more hexes isn't really a line at all. Each unit could be attacked and surrounded by four or more attackers without hindrance from the nearest other defender or its zone of control.

Sometimes the overall situation requires a defensive line arranged "against the grain" (Fig. 5–6). A common result is an analogue of the "with-the-grain" line.

If the defensive line must be made up of quite dissimilar units (or if stacking is allowed but there aren't enough units to allow a pair of units on each position), it's often

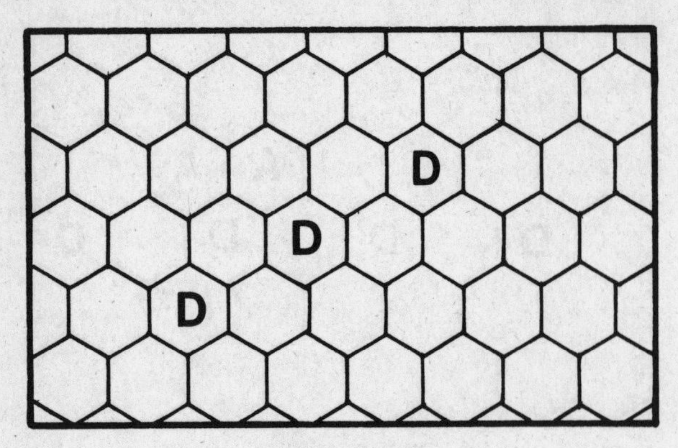

Fig. 5–6

better to have a zigzag line, as shown in Fig. 5–7. The positions in front (marked S), each of which can be attacked from three hexes (marked *A*), should be occupied by the strongest of the defender's units (more than one, if stacking is allowed). The rearmost positions (marked *W*) may safely be garrisoned by a weaker force, since they can only be attacked from a single hex (marked *a*).

Further advice on tactics and strategy in wargaming can be found in *The PLAYBOY Winner's Guide to Board Games* by Jon Freeman and in the *Comprehensive Guide to Board Wargaming* by Nicholas Palmer. Suggested tactics for specific games can also be found in issues of *The General, The Space Gamer*, and other magazines.

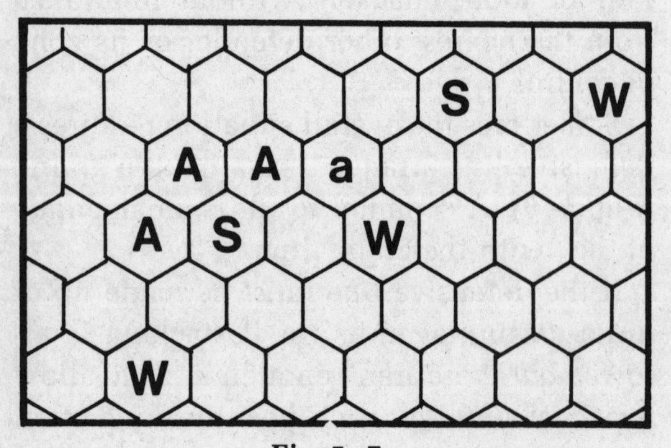

Fig. 5–7

WHAT THE LEFT HAND IS DOING: SOLITAIRE PLAY

Due to the time requirements and a dearth of available partners, most wargames end up being played by a sole person. The most common method of solitaire play is the unplanned or step-by-step approach. For example, you might study the situation and make what appears to be Wellington's best moves. Then you go to the other side of the board (literally or just mentally), study the new situation, and make what look like the best moves for Napoleon's units. Then it's back to the Iron Duke again. The biggest difficulty with this approach is that it is *s-l-o-w*. Surprises are not possible, and your impression of Wellington's best moves is colored by your knowledge of Napoleon's intentions. Like a game that is too well balanced, this equality of knowledge and tactics often produces a stalemate.

The most obvious alternative is to follow history. Wherever possible, have the units do whatever their counterparts did in real life. Gaps in knowledge can be filled in with common sense and some imagination. This method allows you to see how and why the original campaign succeeded—or where it broke down. It also allows you to evaluate how accurately the game captures the facts and the flavor of the actual battle.

This historical method is not without drawbacks, either. For one thing, some generalized games like *Squad Leader* and *PanzerBlitz* have no specific historical precedent to follow. For another, despite your intentions, the results may diverge from history: "Pickett's charge" may sweep everything before it; Blücher's Prussians may stay lost; the Alamo may stubbornly hold out against all odds. Then what do you do?

More to the point, this is only a temporary solution because repeatedly replaying the Battle of Waterloo according to history gets as boring as watching the same toilet paper commercial every hour on the hour.

The intuitive or "blitzkrieg" approach to solitaire play, which can be combined with the historical reply if you're not fussy about precise details, has the virtue of being fast. You look at the basic situation at the beginning of the game and determine a general plan of action for both sides. In *Kassala*, for instance, you might decide that the Moslems will attack the town of Udaka first and then work their way down the flank of the Christian positions until they get to Kassala. The 3·2 infantry units will take the west flank, behind the wadi; the 4·2's will go down the center; and the cavalry will advance into the central plain and attack across the trenches. The Christians will hold their positions, falling back slowly toward Kassala if necessary, and, if they lose Udaka, they won't try to retake it until late in the game, when the Moslems have lost their numerical advantage.

Simple, isn't it?

To carry this sort of blitzkrieg through properly, you can't spend too much time counting various factors and endlessly planning attacks to get the best possible odds. It's not necessary to go all out to "win" when you can't lose. This also makes it easy to simulate surprise, ignorance, blunders, or commanders with one-track minds. With practice, you can learn to grasp the essentials of a situation quickly, and you can get a good feel for situational tactics. More important, perhaps, you get to play twice as many games this way. This allows you to discard unsatisfactory games more rapidly and spend more time with the ones you really like.

PART 2

Evaluating the Wargames

IN THE rapidly changing world of wargaming, it isn't possible to review all the games in print. Instead, we have attempted to give you a representative selection of games that are generally available: good ones and bad ones, "classics" and "young Turks," games to avoid and games you shouldn't miss. Although we have given our opinion of each game's merits, we have also attempted to include enough information so that you can make up your own mind. If the subject, period, or slant is personally appealing, even a not-very-successful game may be of interest.

The following chapters are basically arranged in chronological order, beginning with games set in ancient civilizations and going all the way through modern warfare and into science-fiction, fantasy, role-playing, and computer wargames. Following introductions to each classification are the evaluations of specific wargames.

The format of these evaluations was designed for easy comprehension and comparison. Most terms are self-explanatory, but some points should be clarified at the outset.

The date given with each game title is that of the most recent edition; where relevant, the original publication date may be included as well. In case you can't find a particular game at your local game, hobby, or discount store, the addresses of the publishers are listed in the Appendix. The suggested retail prices of games were the most recent available but are subject to change by the publisher. And some games may be discontinued by publishers or revised without warning. Next to the game price is information about how the game is packaged.

For many gamers, *Playing Time* may be the single most important entry; it doesn't include time spent learning the game, which, without an experienced player to help, will require an additional period of between fifteen minutes and several hours, depending primarily on the complexity of the game. *Balance* covers not only equality of strength and chance at victory, but also, where appropriate, comparative interest and enjoyment.

The games have been given "capsule" evaluations in six areas:

Presentation includes the physical quality of the game's components, graphic appeal, and thoughtfulness of layout and organization.

Rules covers both their completeness and clarity. Are the rules understandable? Do they tell you all you need to know to play the game?

Playability is not intended to be solely a measure of simplicity; some games are simple but don't "work" as games. Imagine a variation of checkers with 5,000 pieces on a forty-foot-square board. Time, convenience, and simplicity are all factors that go into answering the question: How much trouble is it for two people to sit down and play the game?

Realism rates how realistically a game depicts its subject. This may include the accuracy of the map and orders of battle, the balance of forces presented, the appropriateness of period tactics, and the general "feel" of the subject. For science-fiction and fantasy games, this term is extended to encompass believability and verisimilitude. Is the game a faithful treatment of its subject? Is it reasonable? Is it convincing?

Complexity is equally straightforward, though it should be noted that the number of rules is more important in this rating than the sheer number of units or playing pieces, although this is also a factor. As a result, some games with very few units, like *Ogre*, or *Richthofen's War*, may be somewhat easier to play or more sophisticated—or both—than their complexity ratings alone might indicate. Complexity ratings range from 1 (extremely simple) to 10 (extraordinarily complex).

Finally, the *Overall Evaluation* for each game is *not* specifically tied to the other five ratings, since the whole may be more or less than the sum of its parts. Nor are those five aspects, however relevant, the only ways to look at a game. It does, however, attempt to evaluate whether the game is worth your time and money.

CHAPTER 6

Remembering Babylon: Games of the Ancient Period

THE GREAT difficulty in simulating or studying the warfare of the first part of the ancient period is that so little is known about it. The Greek historians Herodotus and Thucydides lived during the Golden Age of Greece in the fifth century B.C.—just in time to record the events of the Persian and Peloponnesian wars. Prior to that, however, there are only archaeological fragments and sources generally considered questionable, at least as regards military history.

From recovered household records, king lists, and bits and pieces of correspondence, we can trace with some certainty the broad movements of the empires of the Near East: the rise and fall of the Babylonians, the Assyrians, the Hittites, the Persians, the Egyptians, and others. *Ancient Conquest* nicely re-creates this ebb and flow across the sands of history, but the game makes no attempt to capture anything but the big picture because the details are not known. We know many of these peoples used chariots for war, but we're not sure how. We know that the Hittites had a secret weapon—iron—that enabled them, temporarily, to dominate the area, but we know nothing of their tactics.

The Bible tells us that Joshua won because God made the sun stand still and brought down the walls of Jericho; the tribes of Israel won their battles when God was on their side and lost when God was mad at them. Taken literally, this would mean that a game based on biblical battles is impossible, because any alternative outcome is impossible. A less fundamentalist view must simply acknowledge that the Bible was concerned with the religious and moral aspects of a conflict—not its military particulars. The forces, objectives, and even the battles of *Chariot* are necessarily conjectural.

Homer was far more concerned with military details, but his gods are as much involved as Joshua's Jehovah. Due to Schliemann's persistent pursuit of what "everyone" knew to be no more than a dream, we know that ancient Troy was no myth, and the archaeological evidence of the last several decades strongly indicates that Homer did not fabricate his incredible tapestry out of whole cloth. A simulation of the glorious siege of Troy without the Greek gods would be colorless; without the heroes it would be nonexistent. *Troy* and *Iliad*—the game, not the poem—incorporate both, and, in fact, *Iliad* dispenses with "catapult fodder" entirely. Since all the important combat was between individuals, and since even Achilles was not two or three times as strong as his chief opponents, a differential Combat Results Table is more appropriate than the more usual odds/ratio variety. (*Troy* combines the two methods to re-create the wildly fluctuating fortunes of war recorded by Homer.)

The dominance of the individual in Homeric combat and in that of the ancient period generally can be explained by the physical and economic facts of life. Real wealth is basically a product of technology; theirs was primitive, and their population base was small. Armor was handmade and expensive. Horses were uncommon and expensive. Even without push-button windows and steel-belted radials, chariots were very expensive. Homer's epic makes sense only if you realize that a decent suit of armor, possibly no more than a helmet and a shield of layered and reinforced bullhide, was worth a king's or at least a prince's ransom. Standing armies were affordable only by aggressive empires, and most of the great armies of antiquity would be lost in a contemporary football stadium.

Anyone can point a gun and kill someone. Fighting with a spear or sword and shield, however, takes practice, strength, endurance, coordination, skill, and a particular kind of courage not often demanded or found on a modern battlefield. There is reason to believe that prior to Philip of Macedon and his son Alexander, most land battles were not won by superior generalship but by the side with the better equipped, more determined, more skilled individuals. Sparta was the dominant military power of its time largely because every free adult male Spartan was first and foremost a fighting man with as much individual skill, courage, and discipline as a Sioux or Apache warrior. Athens, whose physically fit but less single-minded inhabitants wrote plays, discussed philosophy, and invented geometry, could compete militarily only by virtue of its navy, itself a function of wealth, technology, and intelligence rather than physical conditioning and skill.

Realistically simulating combat on a genuine man-to-man level is difficult even in a role-playing game; it's next to impossible in the context of a traditional board wargame. The biggest problem with all ancient-period games is that the process of abstraction required to treat the subject at all almost inevitably takes out much of the color or feel of the period.

The Alexandrian and post-Alexandrian periods are more popular than the ages that preceded them partly because we are on somewhat surer ground historically and partly because we can discern in them a real element of strategy and tactics. Caesar and Alexander not only inspired their men; they also directed them. This allows a more meaningful game because it gives the players a legitimate funtion—simulating the commanders.

Philip of Macedon was a tactical genius. His son was likely that and more—a charismatic strategist the likes of which had not been seen before. The father fashioned a new "weapon," and with it Alexander the Great conquered most of the known world.

The weapon was twofold. Strategically, the Greeks had always been their own worst enemy. Philip solved that by conquering Greece and uniting the Greeks. They had always been disciplined individuals. The storied march of the ten thousand had been made across thousands of miles of unknown enemy territory against fearful odds—and without any of their original leaders. Philip welded them into a disciplined unit and armed them with the *sarissa*, a weapon twice as long as their traditional six- to eight-foot spear. The result was a close array of troops fronted by large overlapping shields and bristling with three rows of spear points—the nearly invulnerable phalanx.

The phalanx was an ideal immovable object, but that was also a problem. It wasn't maneuverable; it could neither turn nor pursue rapidly. It needed an irresistible force to go with it, a hammer for the anvil. Philip's hammer was the strongest heavy cavalry of the time.

Tactics were simple. The phalanx met head-on anything and everything the Persians hurled against it. While the phalanx held its position, the cavalry drove off the lighter enemy cavalry, preventing an attack on the vulnerable rear of the phalanx, and then swept into the rear and flank of the enemy infantry, which then either scattered in disarray or stayed and were smashed between hammer and anvil.

A game like *Alexander the Great* should treat not only the physical aspects of ancient combat—facing, movement, strength—but also the equally important intangible factors of morale, discipline, and inspiration. Games that don't have this lack "period feel" and will be hopelessly inadequate as historical simulations.

Elephants, which were encountered later in Alexander's campaign, became a staple in the armies of his successors—and so much so in simulations of this period that one sometimes gets the feeling that the great pachyderms are a game's only reason for being.

The three-cornered conflict that erupted after Alexander's death came to an end with the rise of Rome, the new ruler of the ancient world. Initially, at least, the military power of Rome, like that of the Greek city-states, was based on the citizen-soldier. The centurions might be professionals, but the average volunteers and the generals were not. And at least until the Empire split, the Roman legion was a unit of infantry. Cavalry was barbaric, undisciplined, and unmanly. Even when employed by Rome, cavalry was treated as an auxiliary force and generally expendable.

The legions had four advantages over their opposition: equipment, training and discipline, organization, and capable generals. The Romans had quite good armor—much better than the Gauls and the Britons, who often wore none at all—and carried a heavy, curved, rectangular shield. The legionary had two weapons, a shortsword and a pike (the *pilum*). Both were designed for close quarters. A volley of the short, heavy spears, hurled from nearly point-blank range, decimated the opposing front rank and left the survivors in confusion and disarray. Before their enemies could reorganize, the Romans charged into them and cut them down with their shortswords. Like Alexander's Macedonians, the disciplined Romans were always outnumbered and always victorious. Or very nearly.

The practical Romans seem to have been born engineers. According to Caesar, at least, they erected elaborate fortifications around their camps every night. While their more lasting achievements—roads, walls, aqueducts, and bridges—have little place in a wargame, their fortifications are the crux of one of the better games of the ancient period, *Caesar*, a simulation of the siege of Alesia at the end of the Gallic Wars.

The rise and decay of empire, the decline

of the citizen-soldier and the rise, first, of the professional and later the out-and-out mercenary, civil wars and military coups, and the barbarian invasions that finished it off—the entire sorry but fascinating tale was the subject of an equally sorry but fascinating game, the notorious *Fall of Rome*, the first wargame designed specifically for solitaire play. The subject is a good one, and it will be interesting to see what Simulations Publications, which is completely redesigning the game, comes up with the second time around. *Decline and Fall*, a four-player game from Wargames Research Group of England, was an attractive treatment of the same topic but has not been generally available in the United States.

Evaluations

Alexander the Great (1975—a third edition of a game first published in 1971 by Guidon Games)

PUBLISHER: Avalon Hill Game Co.

SUGGESTED RETAIL PRICE: $12 (boxed)

SUBJECT: The Battle of Arbela (Gaugamela), 331 B.C. (Persians vs. Macedonians).

PLAYING TIME: A fairly long evening is required.

SCALE: Tactical (battle)-level game with units ranging in size from phalanxes to cavalry detachments. Each hex covers about one hundred yards, and each turn runs about thirty minutes.

SIZE: The 22″ x 28″ mounted game board uses large 25mm (1″) hexes.

BALANCE: The game's balance slightly favors the Macedonian king—as it should—although a Persian player who avoids Darius' mistakes can do very well.

KEY FEATURES: The *Alexander the Great* system is not as outdated as its age might indicate. The counters are of three different sizes to distinguish large units from smaller ones. Separate rules cover melee, flank, and rear attacks, missile firing, chariot and cavalry charges, and elephants. A sliding scale is used to indicate changing fortunes in the area of morale; the importance of this intangible aspect of war is unusual in a game that is almost ten years old.

COMMENTS: Resembling a lime-gelatin-and-chocolate-pudding parfait, the map for the second (1974) edition was one of the uglier maps around. The newer map looks somewhat better, but it is equally inaccurate: as history and even the designer says, the battlefield at Arbela was completely level—it had been leveled by Darius. Apart from the map, *Alexander the Great* is not a bad game. The big phalanxes maneuver a bit too freely, perhaps, and the battle doesn't have too much period flavor, but it is a reasonably evenly matched engagement. While the morale rules and facing systems are fairly sophisticated, the rest of the game is overly simplified. The big problem is that it is difficult to re-create a battle such as Arbela on a small-unit, small-map scale without washing out most of the interest. Thus, while the player can have fun playing the game, he doesn't learn much of the subject from it.

EVALUATION:

Presentation—Fair to Good (depending on edition)
Rules—Fair
Playability—Good
Realism—Good
Complexity—5

OVERALL EVALUATION: Good

Ancient Conquest (1975)

PUBLISHER: Excalibre Games, Inc.

SUGGESTED RETAIL PRICE: $9.95 (boxed)

SUBJECT: The political and military—as well as biblical—history of the ancient Near East.

PLAYING TIME: Depending on the number of people involved, it should take about six hours to play.

SCALE: This is a strategic-level game using tribes, "peoples," and groups of infantry/cavalry. The scale is about fifteen miles per hex, and each turn seems to represent generations.

SIZE: The map is 18″ x 24″ with standard 16mm (⅝″) hexes.

BALANCE: As in most multiplayer games, balance depends on the players. Each "side" appears to have an equal chance, despite a wide variance in setups and objectives.

KEY FEATURES: Much of *Ancient Conquest* is quite simple, especially the combat and movement systems. However, the use of random events, auguries, omens ("Your liver has shone, your fate is your own," and other such queasy revelations), and objectives particular to each tribe/kingdom is almost unique in a game even a few years old.

COMMENTS: This is a game that is hard to pin down. It's essentially a multiplayer game without all that much player interaction. *Ancient Conquest* relies heavily on "chrome"—unusual rules and phrases—to lift the game from its basically simplistic and unassuming level. To an extent, this works: parts of the game are rather witty (combat losses are in "smotes"), but some are a bit arch. Ultimately, the amount of fun you get out of this depends on your willingness to enjoy the approach and the subject. It can be interesting, and it certainly isn't stereotyped. *Ancient Conquest* is an unusual game for people seeking an item a bit out of the ordinary.

EVALUATION:

Presentation—Good
Rules—Very Good
Playability—Excellent
Realism—Good
Complexity—4

OVERALL EVALUATION: Good

The Battle of Raphia, 217 B.C.

PUBLISHER: Game Designers' Workshop

SUGGESTED RETAIL PRICE: $5 (resealable plastic bag)

SUBJECT: The land battle between the Ptolemaic Kingdom of Egypt and the Seleucid Kingdom of Syria, 217 B.C.

PLAYING TIME: About two hours are required.

SCALE: In this battle-level (quasi-tactical) game, each counter represents from one hundred to fifteen hundred men plus elephants. The hexes cover about one hundred yards each, and a turn is about thirty minutes.

SIZE: The 18″ x 24″ game map uses 19mm (¾″) hexes.

BALANCE: The weight of the disputed Heavy Madedonian Phalanx units throws the balance in favor of Ptolemy (Egypt), but there is an optional rule to counterbalance this effect.

KEY FEATURES: *The Battle of Raphia* is fairly representative of ancient-period battle games in its use of facing, the effect of leaders, morale, and weapons differentiation. Unusually, however, there is no missile combat—but then there seem to be no "missile" units present. Of course, elephants are given special rules. (Ah, where would ancient games be without elephants?)

COMMENTS: *The Battle of Raphia* was one of the earliest in GDW's Series 120 games, designed for speed (a target playing time of 120 minutes) and ease of play rather than historicity. Thus, there is little to relieve an otherwise standard situation. Ignoring most of the other terrain, both sides set up facing each other a few hexes apart in the center of the map, and then it's hack and slash, as two lines of units dice each other to death. For those who are not interested in moving around and who prefer to get right down to the bloodbath, this could be of some interest. It *is* easy to play.

EVALUATION:
- Presentation—Good
- Rules—Very Good
- Playability—Very Good
- Realism—Fair
- Complexity—4

OVERALL EVALUATION: Fair

Caesar (1976—originally published in somewhat different form on an amateur level in 1970 as *Alesia*)

PUBLISHER: Avalon Hill Game Co.

SUGGESTED RETAIL PRICE: $12 (boxed)

SUBJECT: The game pits Caesar against Vercingetorix in the epic siege of Alesia, 51 B.C., the culmination of the Gallic Wars.

PLAYING TIME: Up to eight hours are needed for a complete game.

SCALE: This is a tactical siege game that feels more "operational" than anything else. Roman units are mostly in cohorts; the Gauls are in groups from each tribe.

SIZE: The large 28″ x 33″ game board uses 16mm (⅝″) hexes. You may need extra space to play this.

BALANCE: This is a nicely balanced game.

KEY FEATURES: There is very little that is unusual or different in the mechanics of this game; almost all the rules systems have been used elsewhere. Movement and combat—both missile and melee—are so commonplace as to seem almost banal. Despite the quasi-tactical level of the simulation, all units have standard combat and movement factors. The fact that defensive fire takes place *during* movement is interesting. The special rules for fortifications and siege towers, off-board Gallic movement, and the optional reconnaissance rule are all well done.

COMMENTS: *Caesar* (nee *Alesia*) was one of the few legends in wargaming. Copies of the original game are almost impossible to obtain, and the publication of the game by Avalon Hill was a big day for students of the period. Nor were they disappointed: *Caesar* is one of the better games around. The siege of Alesia, a triple encirclement, is one of the most interesting

Caesar, *from Avalon Hill, is based on the Battle of Alesia.* (PHOTO: AVALON HILL GAME CO.)

military situations extant. By not encumbering the game with a horde of complex rules, the designer produced a slam-bang game of feint, surprise, and sudden combat. Both players must make tough, nail-biting decisions from beginning to end. The research on the Gallic tribes is excellent. The game's only drawback, aside, perhaps, from its size, is that the emphasis on surprise, which makes it such a contest for two players, renders it somewhat ill suited for solitaire play.

EVALUATION:
Presentation—Good
Rules—Good
Playability—Very Good
Realism—Good
Complexity—6

OVERALL EVALUATION: Very Good

Caesar's Legions (1975—a revised version of *Eagles,* published by Game Designers' Workshop in 1973)
PUBLISHER: Avalon Hill Game Co.
SUGGESTED RETAIL PRICE: $12 (boxed)
SUBJECT: Various Roman campaigns near and across the Rhine River, from 58 B.C.

to A.D. 68. Some naval rules cover river and sea movement.

PLAYING TIME: This game takes three to six hours, depending on the scenario.

SCALE: This is an operational-level game, with units of varying sizes (from cohort to legion). Each hex covers about ten miles of territory, and each turn runs about two days.

SIZE: The 22″ x 27″ game board uses 16mm (5/8″) hexes.

BALANCE: The Germans have a tough time denying the Romans, but it can be done. With uneven players, the better one should take the German side in most scenarios.

KEY FEATURES: For reasons that are more than a little obscure, considering the absence of similarities in the two periods, the combat system of the original *Eagles* game was discarded in favor of the system used in *1776*. The tactical cards from *1776* are also used, and they are even less satisfactory here than they were before. Standard rules cover the use of large units and their breakdown into smaller ones. Each scenario adds special considerations, such as forts, ambushes, desertions, and the all-important mobilization rules. Most of the emphasis is on the extraneous material, and the basic system has been kept simple.

COMMENTS: When Avalon Hill bought this game from GDW, they immediately corrected the glaring mistake of the earlier version by adding an overrun rule. They then proceeded to add a great deal of unnecessary junk. They did make the game quite colorful, but it is now somewhat overdone. There are only two scenarios that matter, both of which concern the loss and regaining of Varus' Eagles (the standards of a Roman legion). Aside from being dull, the two Julius Caesar scenarios are simplistic and ahistorical. The Batavian Revolt isn't bad, but it has little historical impact. Even the main scenario, the campaign of Drusus Germanicus to regain the Eagles from Arminius, is simply a combination of hide-and-seek and ring-a-levio. It's a lot of fire and effort, simulating nothing.

EVALUATION:
Presentation—Very Good
Rules—Very Good
Playability—Very Good
Realism—Fair
Complexity—5

OVERALL EVALUATION: Fair

Chariot (1975—a complete remake of *Armageddon*, published in 1972)

PUBLISHER: Simulations Publications, Inc.

SUGGESTED RETAIL PRICE: $12 (plastic box; also available, with *Spartan, Legion, Viking,* and *Yeoman* in $35 PRESTAGS Master Pack)

SUBJECT: The game covers a large number of land battles, both actual and mythical, that occurred during the "Biblical Age," 3000–500 B.C., including Megiddo, Kadesh, and others.

PLAYING TIME: It takes two hours for most scenarios.

SCALE: A tactical-level game using counters representing one hundred to one hundred twenty-five men and/or horses. Each hex covers fifty meters and each game turn represents five minutes of real time.

SIZE: The 23″ x 35″ game map uses 16mm (5/8″) hexes.

BALANCE: Some scenarios are better balanced than others, but there are enough evenly matched battles to satisfy anyone.

KEY FEATURES: *Chariot* is the first game in SPI's PRESTAGS (*Pre-Seventeenth Century Tactical Game System*) series, all of which (*Chariot, Spartan, Legion, Viking,* and *Yeoman*) use the same basic rules while adding special ones for the peculiarities of the warfare of their era. One map (different in each game) with a wide variety of terrain is used as a standard, if ahistorical, backdrop for the ten or more scenario/battles given. Many weapon types are available—bowmen, heavy cavalry, spearmen, etc.—each of which has different capabilities and strengths. The system is the standard for most tactical battle games of the pregunpowder era: Fire—Move—Melee. Optional rules bring in facing, panic levels, and possible simultaneous movement. Leaders play an important part in controlling the armies.

COMMENTS: The PRESTAGS system is a marvelous, if colorless, way to introduce people to wargaming and the early historical periods. Once the rather simple standardized rules have been learned, there are almost a hundred different battles immediately available in the system. This multiscenario approach to games has recently fallen into disfavor in some quarters, however, for several reasons. Neither

SPI's PRESTAGS MasterPack with the games Chariot, Spartan, Legion, Viking, *and* Yeoman. (PHOTO: SIMULATIONS PUBLICATIONS, INC.)

the maps nor the counters have historical designations. Thus, there is little color, and player identification is minimal; one battle has little to distinguish it from another. Moreover, because of counter-mix limitations and the generalized nature of the system, many of the orders of battle for the individual scenarios are warped or downright inaccurate. The most conspicuous hole in the system, which gives only marginal insight into actual pregunpowder tactics, is a lack of morale rules. On the other hand, the system is clean, and most battles can be played in a short time. If you are not too concerned with the limitations of the system, the variety of scenarios makes each game a bargain.

EVALUATION:
Presentation—Good
Rules—Very Good
Playability—Very Good
Realism—Fair
Complexity—5

OVERALL EVALUATION: Good

The Fall of Rome (1973)

PUBLISHER: Simulations Publications, Inc.

SUGGESTED RETAIL PRICE: $9 (resealable plastic bag)

SUBJECT: The decline of the Roman Empire and the barbarian invasions, from A.D. 100 to 500 (land only).

PLAYING TIME: It takes four to five hours for each of the six scenarios.

SCALE: The strategic-level game uses strength points of men, representing ten thousand to fifty thousand men each. Each turn is one year, and the small map covers all of Europe.

SIZE: The 17″ x 22″ map uses areas—provinces—rather than hexes.

BALANCE: Balance is immaterial, as the game is for solitaire play only. Therefore, you can't lose (then again, you can't win, either).

KEY FEATURES: The key point is that the game is designed for solitaire play. Thus, the movement of the non-Roman forces is rigidly—if not too clearly—regulated by a system of reaction to the freer Roman movement. Combat is fairly standard odds/ratio. There are tables covering internal revolts, creation of barbarian hordes, and such, and rules concerning treasuries, the raising of militia troops, who controls whom (and when)—plus a myriad of other estoric rules, all marvelously and comically obscure.

COMMENTS: The legendary *The Fall of Rome* is wargaming's version of the 1962 Mets. Particularly given the resources of SPI and the minds devoted to this subject, it is probably the worst game ever published on an actual historical subject. The rules are an abysmal, if somewhat humorous, disaster; the errata sheet for the game is longer than the original rules—which, in itself, creates as many problems as it solves. Even if you can figure it out, the game system requires so much referring and page-turning just to see what comes next that most players will either be content to let the barbarians take over or play Nero and burn the whole damn thing to avoid confusion. An unmitigated disaster.

EVALUATION:
Presentation—Fair
Rules—Abysmal *
Playability—Poor
Realism—Fair
Complexity—6

OVERALL EVALUATION: Very Poor

* Our lowest rating (Very Poor) was inadequate in this case.

Iliad: The Siege of Troy (1978)
PUBLISHER: Conflict Game Co.
SUGGESTED RETAIL PRICE: $11.98 (boxed)
SUBJECT: The great ancient/mythological siege of Troy. The emphasis is on land conflict; there are some peripheral naval rules.
PLAYING TIME: About three hours are needed.
SCALE: This is a tactical game with operational overtones; the counters represent the individual participants (Paris, Achilles, Hector, etc.). Each hex represents about one mile, but the turn scale, nominally one year per turn, is quite abstract.
SIZE: The mounted 19″ x 25″ game board uses 19mm (¾″) hexes.
BALANCE: The Greeks don't have the wooden horse, but they do have Achilles and a slightly better chance of winning—very slight, though, and not enough to upset anyone.
KEY FEATURES: The heart of this game system is the method of individual heroic combat, in which each hero is rated for attack, defense, and divine favor—the ability to be saved by the gods from death or defeat. The differential CRT results in wounds or death to one or the other of the combatants. Since Achilles is, appropriately, almost invincible, there are special rules limiting his use. Astride the conflict (and the basic system) are the gods, to whom the players may appeal for special favors. Random events are added by a special decks of cards.
COMMENTS: While the game is hardly complex, it is an effective simulation of the complex (divine) relationships and events found in Homer. In this respect, it works much better than *Troy*. The random events for the gods are colorfully and interestingly treated, and the comparative strengths of the heroes are acceptable (even if the game, like *Troy*, persists in treating "Ajax the Lesser" and Teucer as two different persons—a point questioned by modern scholarship). Much of the action is quite stereotyped, and if the Greek player is careful, there isn't a great deal for the Trojan to do; after all, he is under siege. Gaming a piece of fiction (a reasonable way of viewing this) can be a difficult process (see *War of the Ring*), and the company seems to have opted for providing enjoyment rather than slavishly sticking to detail. On that level, it has succeeded rather well.
EVALUATION:
Presentation—Good
Rules—Good
Playability—Very Good
Realism—Good
Complexity—5
OVERALL EVALUATION: Good

Legion (1975—a remake of *Centurion*, published in 1971)
PUBLISHER: Simulations Publications, Inc.
SUGGESTED RETAIL PRICE: $12 (plastic box)
SUBJECT: Tactical land warfare in the Roman Age, 100 B.C. to A.D. 700. Battles included as scenarios are Carrhae, Pharsalus, Teutoburger Wald, Placentia, Adrianople, and others.
COMMENTS: This game is part of SPI's PRESTAGS series. See *Chariot* for a review of that system.

The Peloponnesian War (1977)
PUBLISHER: Simulations Canada
SUGGESTED RETAIL PRICE: $10.99 (resealable plastic bag)

SUBJECT: The war between Athens and Sparta on land and sea, 431–404 B.C.

PLAYING TIME: It takes eight hours or more for a complete game.

SCALE: This is a strategic-level game that uses large groups of infantry and cavalry and fleets of triremes. Each turn is a year, and each hex covers about ten miles.

SIZE: the 22″ x 28″ game map uses 16mm (5/8″) hexes.

BALANCE: In this game the Spartans have a slight edge in strength, but an Athenian player who knows how to use his navy can more than make up for this difference.

KEY FEATURES: There is a strong emphasis on the logistics of warfare, which is unusual for an ancient campaign game. SPI's *The Conquerors* is the only other such game to concentrate on this aspect. Thus, there are rules for mobilization and funding, as well as normal supply rules involving control of cities and naval pipelines. The CRT reflects units lost to both attacker and defender, and there are rules covering morale. Some of the famous leaders, such as Pericles and the infamous Alcibiades (whose turncoat machinations are reminiscent of a Gilbert and Sullivan operetta), have their own counters.

COMMENTS: *The Peloponnesian War* has several items weighing heavily against its success. The counters are garish and hard to read, and the map colors would induce nausea in a veteran nurse. In addition, the game was apparently designed and "developed" (a little) in only two months. Thus, the interesting, if somewhat loosely structured, situation is simulated with simplistic rules and careless design work —hardly a recommendation for a fun evening. Yet the situation itself almost—but not quite—redeems the game from perdition (if not oblivion). The game's designer does have, in general terms, a vague handle on the feel of the war, but the subject deserves better.

EVALUATION:
Presentation—Poor
Rules—Fair
Playability—Good
Realism—Fair
Complexity—5

OVERALL EVALUATION: Fair

Pharsalus (1977)

PUBLISHER: Game Designers' Workshop

SUGGESTED RETAIL PRICE: $9 (resealable plastic bag)

SUBJECT: The land battle between Caesar and Pompey during the Roman Civil War, 48 B.C.

PLAYING TIME: This depends on how cautious the players are; two wily opponents could use up the better part of six to seven hours.

SCALE: This is a tactical-level battle game, with each counter representing a cohort. Each hex covers forty yards, and each pair of (alternating) turns encompasses six minutes of real time.

SIZE: The 22″ x 28″ game map uses 16mm (5/8″) hexes.

BALANCE: The game would tend to favor Pompey—and his larger army—if it weren't for some subtle stacking rules, rules so subtle we missed them the first two times through (a comment either on our grade of play or the capability of the rules writer—or both). With those rules it's a pretty nip-and-tuck affair.

KEY FEATURES: The somewhat interwoven sequence of play reflects a greater tactical orientation that is typical of battle games of this period. Units have facings depending on their weaponry; fatigue is a factor (a major one); and there are special rules for units with *pila*. And, as is usual with GDW games, morale plays a large part. Oh yes, that overlooked stacking rule: Only a certain number of strength points may melee, regardless of the size of the unit. This is a nice rule that rewards the size of a unit not with combat capability directly but rather with staying power.

COMMENTS: Despite much chromatic machination, most ancient battle games tend to look and feel alike. As a consequence, they are heavily dependent on the interest of the subject. This game's system is somewhat more unusual than others—a step above the straight SPI PRESTAGS system—and the battle itself is an interesting one. Nonetheless, the dull, almost nonexistent terrain serves to define the battlefield rather than decorate or influence it, and like all ancient battles, the affair is limited in maneuver and heavy on shock. Within these limitations, *Pharsalus* does quite well: you even get the feel of moving individual legions, withdrawing tired troops, and gearing up for the big charge, as opposed to just pushing counters around. A nice effort.

EVALUATION:
Presentation—Good
Rules—Good
Playability—Good
Realism—Good
Complexity—6

OVERALL EVALUATION: Good

The Siege of Jerusalem, 70 A.D. (1976)

PUBLISHER: Historical Perspectives

SUGGESTED RETAIL PRICE: $15 (unpackaged except for shipping box)

SUBJECT: The Jewish Revolt in Palestine against the Roman Empire.

PLAYING TIME: A full game of the entire siege will take about ten to twelve hours, but the smaller scenarios can be done in three to four hours.

SCALE: This is a tactical-level siege game with cohort-sized Roman units and small groups of Jewish soldiers roughly equivalent, perhaps, to companies. Each turn represents about an hour of time; each hex is equivalent to about one hundred yards.

SIZE: The 29″ x 45″ game map is cut into four sections and requires a fairly spacious playing surface. The standard 16mm (5/8″) hex is used.

BALANCE: The game is fairly even in most scenarios.

KEY FEATURES: The system, which is quite similar to Avalon Hill's *Caesar*, uses an extended form of "Assault Period," in which the four Roman legions may come on any time anywhere on the map, but they may not stay for more than eight turns due to fatigue. The sequence of play is an interwoven one of fire, movement, and melee, with strengths based on range and target "cover." A wide variety of siege equipment is used to batter, ram, and smash the walls of the city, after which come the brutal assaults with ladders and ramps. Movement inside and outside the city is handled so that the players are always racing units from one section to another in a mad effort to strengthen their positions. Other rules cover fatigue, bivouacking, and replacements.

COMMENTS: *The Seige of Jerusalem* is one of the finest amateur games ever published. Aside from the usual physical drawbacks of nonprofessional games (here kept to a minimum), this is a masterful job from all angles. In terms of player excitement, it is almost as good as *Caesar*, which it resembles in situation as well as system, and head and shoulders above many other siege games. What makes it work is that the designers have abstracted a greal deal of the tedious "wall-busting" and concentrated on the bloody assaults. Both players fight the clock as well as each other. Some of the rules are a bit hazy, and a stronger sense of organization might have helped. But this is a game that was not only designed well but developed well, and it shows in the level of excitement it provides.

EVALUATION:
Presentation—Good
Rules—Fair
Playability—Good
Realism—Very Good
Complexity—7

OVERALL EVALUATION: Very Good

Spartan (1975—a reworking of *Phalanx*, originally published in 1971)

PUBLISHER: Simulations Publications, Inc.

SUGGESTED RETAIL PRICE: $12 (plastic box)

SUBJECT: Tactical land warfare in the Hellenistic Age, 500–100 B.C. Battles included as scenarios are Marathon, Chaeronea, Granicus, Issus, Arbela, and others.

COMMENTS: This game is part of the SPI PRESTAGS system. See *Chariot* for a review of this system.

Troy (1977)

PUBLISHER: The Chaosium

SUGGESTED RETAIL PRICE: $10 (resealable plastic bag)

SUBJECT: Covers the ancient siege of Troy on land; some sea rules.

PLAYING TIME: It takes two to four hours, depending on the scenario chosen.

SCALE: This operational-level game has tactical overtones; it features individual heroes leading groups of men. The map uses a scale of three hundred meters to the hex, and the game turns represent four hours. (The periods of inactivity are ignored.)

SIZE: The 23″ x 34″ game map uses 16mm (5/8″) hexes.

BALANCE: Most of the scenarios seem relatively well balanced.

KEY FEATURES: The emphasis in *Troy* is on "introducing" as much of history and the Homeric saga as possible. Considering this intent, the rules are remarkably simple—even simplistic. Movement and combat are about as simply handled as any game on the market: you simply move the unit, compare factors, and check results. Most of the rules concentrate on special items: siege, supply and booty, and the deities. Each player may use one of a series of cards to invoke a god(dess) and get his/her particular form of aid. The special rules are introduced gradually in scenarios that cover all the (presumed) wars around Troy—not just the famous siege of Agamemnon and Achilles. There is (inevitably) a wooden horse option.

COMMENTS: Despite the obvious love for and the exceptional insight into the subject (half the forty-eight-page rule book is background on the wars, the participants, etc.), *Troy* is not nearly as successful as its counterpart, *Iliad*. The reasons? Graphics, for one. *Troy*'s map (accurate as it may be) and counters are unattractively rendered in bizarre and unappealing pastel shades that make the game painful to contemplate. A more basic reason is that the game is too simple to satisfy anyone except the rank beginner. There is little interaction between the deities, and the basic game system is so easy that most players will quickly lose interest. *Troy* illustrates the difference between input and output: what came out was not worthy of the information and affection that went into it.

EVALUATION:
- Presentation—Fair
- Rules—Very Good
- Playability—Very Good
- Realism—Fair
- Complexity—3

OVERALL EVALUATION: Fair

CHAPTER 7

Medieval Warfare

THE MIDDLE AGES is one of the least popular periods in wargaming—with good reason.

Wargamers look at war, historically, as the ultimate means of settling disputes. They see battle as an exercise in strategy, tactics, and weaponry. These are not balanced or "realistic" views; they are not intended to be. Gamers abhor the total reality of war as much as anyone; instead, they treat it abstractly as a contest, a game—a more colorful chess, a Super Bowl for higher stakes.

Medieval warfare is marvelously unsuited to this gaming point of view. High-level conniving was common enough, but a real grasp of grand strategy was rare, and a sense of proper military strategy was even rarer. Grand tactics were nonexistent, and low-level tactics were crude. Caesar's legions or Alexander's Macedonians could have conquered and united the Europe of A.D. 1000 in a good summer. Military organization had decayed with the Roman Empire; military genius was dormant. Wars, fought by troops raised by levy over issues important to their rulers and almost no one else, seldom achieved lasting results.

In western Europe, able-bodied men owed a period of annual service to the feudal lord, who in turn owed service to the nobles above him in the pecking order, on up to the king (where there was one). Except for metalworkers or other artisans, everyone was expected to turn out for military service as soon as the spring planting was completed. Pentecost Sunday was a convenient date.

Before the Norman Conquest, this was the basis of the English military. Similar approaches were used by the Normans in France, Italy, and Germany, and by the Franks in France and Spain. During the period of service, the nobles would use these "summer soldiers" to continue any unfinished disputes and, usually, to start some new ones.

Through the time of the Crusades, Europe was never at a loss for new invaders. The barbarian invasions, which had finished off the decaying Roman Empire, in later centuries often accomplished little except population shifts that set the stage for future conflicts over land ownership and political suzerainty. Seeking new lands and subjects to tax, kings began hostilities on the most absurd pretexts and justified wars on the basis of family ties and blood relationships that were outrageously obscure when they weren't wholly imaginary. No lord felt good if he undertook to feed his levied troops for several months and didn't get anything out of it except the bill. Little wars were almost an annual occurrence, if for no other purpose than to keep the troops in practice for the next important war.

Beneath all the political machinations of the period lay the unburied ghost of the Roman Empire. Petty kings and princelings wholly unsuited to the task spent most of the Dark Ages trying to rebuild the Roman system and reestablish the central authority of the Empire—with themselves at its head, of course. Compared to the original, the Holy Roman Empire was a joke, but people everywhere took the idea of it seriously.

Some of the most interesting and attractive games based on this era, such as *Kingmaker, Machiavelli,* and *Timbuktu,* are grand-strategy simulations of these internecine struggles for power, land, and recognition. The free-for-all nature of these conflicts makes a good basis for multiplayer games.

There are only two other sorts of simulations of this period: the siege and the ordinary field battle. Naval warfare simply didn't exist; the "warships" of the Vikings were nothing but vehicles of transportation.

The nature of medieval combat was dependent on the economics of the time, which was not so different from that of the Bronze Age. Only a small minority, the knights and nobles, could afford proper equipment, training, and a horse. A peasant levy might possibly wear a metal cap and a leather jerkin but certainly no other bodily protection. Peasant arms were sometimes actual weapons of war but more often, particularly in minor wars and border disputes, they were the same hoes, axes, and scythes used in the fields. These odd weapons were reasonably effective—at least against other peasants—and were necessarily kept sharp and relatively clean for normal use. Nor did they need special training to wield. Finally, of course, they were the only weapons commoners could afford.

Against a charge of armored knights on horseback, peasants armed with farming implements were about as effective as infantry bayonets against a German King Tiger tank. While the Mongols and other lightly armed horse archers dominated the plains of the East in the manner of the Parthians of ancient times, in western Europe cavalry was the armor branch—veritable tanks on the hoof. Until the Battle of Agincourt in 1415, the discrepancy between mounted knights and foot soldiers tended to increase, as over the centuries armor—for those who could afford it and the specially bred giant chargers needed to bear the weight—grew increasingly sophisticated, elaborate, and impenetrable to anything the infantry could do.

The death rate among the nobles—at least in battle—was surprisingly low. This was due partly to the protection of their armor and partly to the fact that anyone who could afford the full panoply of war could afford a ransom. Peasants lacked both and were eminently expendable.

Most bows lacked the range and penetration needed to deter cavalry. Mechanical crossbows could penetrate the heaviest armor, but they were slow and clumsy, and their range was unspectacular. Longbows required considerably more training, but in the hands of skilled bowmen they had an effective range of up to six hundred yards. Their planned and concerted use at Agincourt and other fifteenth-century battlefields brought to an end the reign of cavalry and an elite military nobility as surely as the introduction of firearms and artillery.

The single exception to the dominance of cavalry prior to Agincourt in the Middle

Ages was a dense formation of Swiss pikemen, in effect a descendant of the Macedonian phalanx. This formation was much more maneuverable than its ancient counterpart because no shields were carried. For defense as well as offense, the Swiss relied on two weapons: a long pike much like the *sarissa* and a long, hook-bladed halberd or bill. The former countered the chargers, while the latter were used to unhorse and dispatch the riders. Part of the pikemen's success was due to the mountainous terrain of Switzerland, which was wholly unsuited to mass cavalry charges. Discipline, courage, and spirit were nearly as important, however; over the years, as their confidence grew, the Swiss often chased onto the lowland plains cavalry that had the audacity to attack a Swiss canton. On those relatively rare occasions when fleeing cavalry would try to take advantage of the terrain, reform, and counterattack, the Swiss pikemen simply formed a square—the hedgehog—and withdrew in good order.

The simplest way to translate medieval warfare is fairly obvious. Movement rates, a matter of marching speed, can be determined from historical accounts, as can the effectiveness of different weapons and groups of combatants. Archers can be treated as forms of artillery of varying range and effectiveness. Print the counters and a simple rule book and *voilà:* a medieval wargame.

Some games are little more than that, but there are quite a few difficulties with this method. The most significant level of action—man-to-man—doesn't really survive the basic process of abstraction. Man-to-man fighting can be handled somewhat better in a role-playing game, and, indeed, a pseudo-medieval milieu forms a background common to *Dungeons & Dragons, Chivalry & Sorcery,* and a number of other fantasy role-playing games.

Another problem is that the tactics of the modern era are, to say the least, more developed than the tactics of the time, when a massed frontal assault was considered a clever idea. Thus, fancy tactical maneuvering by a game player is totally unhistorical. One unsatisfactory response to ahistorical tactics is a rule or set of rules that limits a player's actions to those followed by the historical commanders. SPI's *Agincourt* is an extreme example of this approach. Historian-gamers may be satisfied with the result—once or twice, anyway—but eliminating all of a player's options ruins a game for the other types of wargamers.

Finally, cardboard units on a game board respond to orders with one hundred percent reliability, which is about ninety percent better than most medieval commanders could manage. Even before the organization of the Romans and the genius of Alexander, the presence of a leader on the field of battle was an important psychological factor. As long as he was obviously present and involved in the fighting, his men were less likely to succumb to panic. Like Caesar, medieval leaders inspired their troops to stand in the face of adversity, to fight harder, to go "once more unto the breach, dear friends," instead of giving up or retreating. These intangible factors can be handled with leadership, morale, and panic rules like those used in *Alexander the Great* and other games of the last chapter. Such rules may be annoying to the "competitor" and not altogether pleasing to the "assassin" and the general gamer, but they are, perhaps, a reasonable compromise between playability and historicity.

As the castles and fortresses of Europe grew ever more elaborate and difficult to assault, siege tactics developed accordingly. Sieges were an important part of medieval military operations, but they were usually slow, drawn-out affairs whose success turned on such questions as: Will the inhabitants surrender before they starve? Will we finish a tunnel under the walls before the defenders find and collapse it? Will the disease spread by the dead horses catapulted inside leave people too sick to man the walls? Thrilling stuff, that, full of action and excitement.

This significant but formerly neglected area has been the subject of a number of recent games like SPI's *The Art of Siege Warfare* collection, among others. But such games have not, on the whole, been too successful. Sieges simply lack the action and movement wargamers prefer, and the treatment of the esoteric details has been, perhaps inevitably, unsatisfyingly abstract.

The medieval period, however, has the advantage of not being done to death, which is more than can be said for World War II's eastern front or the Battle of Waterloo.

Evaluations

Acre (1978)

PUBLISHER: Simulations Publications, Inc.

SUGGESTED RETAIL PRICE: $9 (resealable plastic bag; also available boxed as part of *The Art of Siege Warfare* for $27)

SUBJECT: The game is based on the siege of Acre by the Crusaders under kings Richard, Phillipe, and Guy, as well as the countersiege operations of Saladin in 1191.

PLAYING TIME: The game takes five to seven hours to play.

SCALE: This is a tactical-level game with some operational overtones. Each unit represents six hundred to eight hundred men, each turn two days, and each hex fifty meters.

SIZE: The 22″ x 34″ game map uses 16mm (5/8″) hexes.

BALANCE: Considering this is a siege, game balance is quite good. This is largely due to the presence of Saladin and his army outside the city.

KEY FEATURES: *Acre* is part of *The Art of Siege Warfare*, but its system was largely inspired by *The Siege of Constantinople*. Since the sequence of play closely follows that of a real siege, most of the rules concern not what to do but rather how to do it. There are Bombardment Tables, Wall Repair Tables, Range Tables, and the standard Fire and Melee Tables—most of them fairly simple. Rules cover almost every phase of a siege and include engineering, subterranean combat, naphtha grenadiers, leadership, towers, and so on.

COMMENTS: Although *Acre* is based on *The Siege of Constantinople*, it is, despite their similarities, not so big a bore as the latter. Since the constant tunneling and bombardment produce no immediate effect, siege warfare is not exactly a thrill a minute. Fortunately, *Acre* shows some interest in the combat aspects of the siege. Thanks to the presence of Saladin's army

The Art of Siege, *a four-game set from SPI, including the battles of Tyre, Acre, Lille, and Sevastopol.* (PHOTO: SIMULATIONS PUBLICATIONS, INC.)

as a constant threat to the besieging Crusaders, *Acre* has a great deal more action as well as a good bit more color—both literally and figuratively—than its parent. As a complete package (with *Lille*, *Sevastopol*, and the unreviewed *Tyre*), *The Art of Siege Warfare* may not have the most playable of systems; it does, however, fill an important historical hole in the hobby. *Acre* is probably the most playable of the four.

EVALUATION:
 Presentation—Very Good
 Rules—Good
 Playability—Good
 Realism—Very Good
 Complexity—6

OVERALL EVALUATION: Good

Agincourt (1978)

PUBLISHER: Simulations Publications, Inc.
SUGGESTED RETAIL PRICE: $10 (boxed)
SUBJECT: The Battle of Agincourt between the English and the French in 1415.
PLAYING TIME: A typical game takes about four hours. (This game was being adapted for home-computer play; when released,

the computer aid should cut game time in half.)

SCALE: This is a tactical game with counters representing from two hundred fifty to one thousand men-at-arms. Each hex is thirty-four yards across, and each game turn represents three minutes of real time.

SIZE: The 22″ x 34″ game map uses 16mm (5/8″) hexes.

BALANCE: This is a big problem because in most games the English will win. In the scenarios in which the French have a chance, the history is so askew that it hurts to play that way.

KEY FEATURES: This is the first game really to treat medieval warfare as a serious subject. The system is intense—down to lost arrows and dead bodies littering the field. Land units are represented by double-sized counters; this requires a new system of movement. Also new is the use of ten- (actually twenty-) sided percentile dice, which allow rolls of 1–100. (New, that is, for board gaming; this has been standard in miniatures and role-playing games for years.) Combat includes both melee and (ranged) fire. Leader combat depends on the stance of the man: standing/parrying, advancing/attacking, and so on. Large-formation melee uses the combat differential cross-referenced to the number of lines (deep) in the formation. There are also rules for stakes, *Fait du Mort* squads, cavalry charges, morale, retrieving arrows, rate of fire, and more. It's quite complex for a medieval battle game, and it adds up to quite an insight into the tactics of the era.

COMMENTS: As a history lesson, *Agincourt* is virtually without peer. As a game, it is a stultifying bore, replete with endless die-rolling and a total lack of heart-stopping action. Two things serve to kill off an interesting system: the situation and the system itself. Given the positions of the units, the result of the battle was a foregone conclusion. Repositioning them would be totally ahistorical, since Henry (the English king) would not have offered combat from a different position. So the French player, with his greatly superior —if somewhat ill-advised—troops, is forced to bang his head against a stone wall. Then there's the system. The designer, James F. Dunnigan, used to have a basic precept for historical simulations: Always supply the illusion of movement. Here he violates his own commands: There is neither illusion nor movement. Players simply roll endlessly to see where arrows are going, where cowards are fleeing, and where the bodies—mostly French ones—are piling up. The result is a history lesson, pure and simple. To save wear and tear on your wrist, read the book.

EVALUATION:
Presentation—Very Good
Rules—Very Good
Playability—Fair
Realism—Excellent
Complexity—7

OVERALL EVALUATION: Fair

The Battle of Agincourt, 1415 A.D. (1978)

PUBLISHER: Game Designers' Workshop

SUGGESTED RETAIL PRICE: $5 (resealable plastic bag)

SUBJECT: The game covers the Battle of Agincourt between the English and the French in 1415.

PLAYING TIME: The game takes two to three hours to play.

SCALE: This is a tactical game, but the exact time and distance scales are not specified. Each counter represents from one hundred fifty to three hundred fifty men.

SIZE: The 17″ x 22″ map uses 20mm (¾″) hexes.

BALANCE: This is reasonable. The numerical superiority of the French is balanced by English advantages in archers, position, and morale and victory conditions that force the French to take the offensive.

KEY FEATURES: Except for morale and facing rules and some minor details, the game uses quite a simple system. The most common result on the odds/ratio Combat Results Table is a morale check. Units that fail this check are routed: they defend at half strength and must retreat if possible. Die roll modifications for combat and morale checks are dependent on leaders and facing. There are no zones of control, and combat is optional. Archers are treated, as usual, as ranged artillery, the English longbows having greater power and range than the French short bows.

COMMENTS: Most of the comments on *The Battle of Raphia*, another Series 120 game, could apply here. The basic situation, facing rules, and fixed starting positions conspire to limit maneuver pretty much to a frontal assault. Besides the subject matter, the only thing this game has in common with SPI's *Agincourt* is a lot of die-rolling. Only the overall circumstances (embodied in the victory conditions) restrict the French side; while the rules encourage the cavalry to charge, they are not required to do so. Since the French, particularly the cavalry, are vulnerable to the English archers, it's a fairly balanced contest. This may not be entirely historical, but it's the only way to make a *game* out of the situation. Unfortunately, there's not enough depth to keep playing it. For beginners—and those who want to teach them—it's not a bad introduction to the basics of combat and morale, which figure so prominently in more complex games.

EVALUATION:
Presentation—Good
Rules—Very Good
Playability—Very Good
Realism—Fair
Complexity—3

OVERALL EVALUATION: For beginners only

Kingmaker (1977—a somewhat different version of a game originally published in England by Philmar in 1974)

PUBLISHER: Avalon Hill Game Co.

SUGGESTED RETAIL PRICE: $12 (boxed)

SUBJECT: This is a multiplayer, diplomacy-oriented game of the fifteenth-century English Wars of the Roses on land and, to a lesser extent, sea.

PLAYING TIME: This is a game for a long evening; since there is no set number of turns, it could last many hours.

SCALE: This is a strategic-level game utilizing individual barons/lords/knights and their retinue. The time scale and hex equivalent are immaterial and abstracted.

SIZE: The 16″ x 22″ game board uses a system of variegated squares and rectangles, resembling an overview of plowed fields, instead of hexes.

BALANCE: This is very good, but subject to fluctuation from game to game.

KEY FEATURES: *Kingmaker*'s system is unique. The main thrust comes from two sets of cards. One set is used to move the action along and supply random events:

Kingmaker, *published by Avalon Hill.* (PHOTO: AVALON HILL GAME CO.)

plague, revolts, storms, and so on. This same set of cards is also used to decide the outcome of the rare battles that occur. The other set of cards represents the lords and barons, as well as their titles and offices, which are distributed randomly throughout the game. Some barons are much better than others. As "pawns," there are the members of the York and Lancaster royal households; the object is to grab one and have him crowned king. The fortunes of a given player fluctuate wildly because of the many random events. The actual mechanics of play are quite simple, although some of the optional rules, such as parliamentary voting, can be a bit obscure.

COMMENTS: *Kingmaker* may be the finest diplomacy-oriented simulation available—even including the hallowed and venerable *Diplomacy*. It allows perhaps as much interplay, while it avoids the stilted and stereotyped play that can afflict the older game's opening moves. The action is furious, and while the game is not a truly accurate portrayal of the Wars of the Roses, *Kingmaker*—far more than most games—imparts a vivid picture of the times. Players quickly find themselves adopting a "baronial" mentality, ruthlessly assassinating dukes because they serve no earthly political good. In addition, the game is exceptionally colorful, especially in its Avalon Hill edition. The rules of the original Philmar edition were somewhat opaque, but while the Avalon Hill map corrects several mistakes in the Philmar version, it is much smaller and definitely inferior to the original. Those who have it should, after making the necessary corrections, use the bigger map. *Kingmaker* won a Charles Roberts Award as Best Strategic Game in 1976 and, more than any other game, proved the effectiveness of using cards to inject flavor and randomness into simulations. Hindered only by the necessity of having at least four players, *Kingmaker* is one of the finest games on the market.

EVALUATION:

Presentation—Very Good (Excellent, with the original mapboard)
Rules—Good
Playability—Very Good
Realism—Very Good
Complexity—6

OVERALL EVALUATION: Excellent

Machiavelli (1978)

PUBLISHER: Battleline Publications

SUGGESTED RETAIL PRICE: $12.95 (boxed)

SUBJECT: This is a power politics game of Renaissance Italy, divided into scenarios covering the period 1385–1530. Land and sea combat are abstractly represented.

PLAYING TIME: This game takes a few hours to forever, depending on player stamina.

SCALE: This is a grand-strategic game using army, fleet, and garrison units. Movement is by area, and each turn represents a season.

SIZE: The 22″ x 27″ game board is divided into irregular areas according to political boundaries (like *Diplomacy*).

BALANCE: The balance is quite good—almost too good, in some scenarios.

KEY FEATURES: *Machiavelli* is, essentially, a very fancy version of *Diplomacy* set in and around Renaissance Italy. The "basic game" is almost indistinguishable from the older game: movement is planned and simultaneous, and combat is resolved by sheer power of numbers. The slight difference is a product of the change in geography; the Italian peninsula places some restraints on movement without naval support, and naval operations assume a premium position. However, where *Diplomacy* remained abstract, *Machiavelli* aspires to historicity. To that end the Advanced Game adds plague, famine, rebellions, assassination, garrisons, loans, bribery—all those rules you've come to know and love with power-politics games. Up to eight people can play, but each of the four scenarios lists the best configuration for use. In all cases, the object is to unify a splintered and diffuse Italy under your control.

COMMENTS: *Machiavelli* is one of the most attractive games available; color is rampant, and the counters, each with the particular emblem of the country involved, are spectacular. After that, however, it's somewhat downhill. Most of the additional rules added to the basic *Diplomacy* system are quite good and impart a good deal of needed flavor. The only major lack is a system for handling religion, a prime motivating factor of the times. Players can take the part of the Vatican or the (Islamic) Turks, but there is no special "feel" for it—the only exception to a treatment that generally gets good marks in that area. The military and political situation is evenly matched; since many of the natural disasters are random, no one can ever be really sure where he will stand. However, a curious result of all this is that in most of the scenarios—especially those with fewer players—an initial period of land-grabbing is followed by endless turns of stalemate, as most countries assume positions of unmovable strength. In essence, the game is *too* well balanced, and only an unusual change in fortune can move it. To a great extent, this was the problem in actuality, but for a game it can be annoying. Yet *Machiavelli* is still good fun.

EVALUATION:
- Presentation—Excellent
- Rules—Very Good
- Playability—Good
- Realism—Fair
- Complexity—6

OVERALL EVALUATION: Good

Robert the Bruce (1978)

PUBLISHER: Fusilier Games

SUGGESTED RETAIL PRICE: $13.99 (boxed)

SUBJECT: The Scots Wars of Independence in the thirteenth and fourteenth centuries, on land and sea.

PLAYING TIME: The game takes fifteen to twenty hours, unless a turn limit is set.

SCALE: This is a strategic-level game, with individual leaders and groups of fighting men. The turns are seasonal, and the hex scale is about ten miles per hex.

SIZE: The 20″ x 22″ mounted game board uses 16mm (5/8″) hexes.

BALANCE: As with most multiplayer games, balance depends largely on the type of players involved. Within this context, the three-player game is the most well balanced.

KEY FEATURES: There are basically five scenarios using from two to five players with the general objective of being crowned King of the Scots. Movement is quite simple, as is combat, which is an unsophisticated odds/ratio, Attacker Eliminated/Defender Eliminated system. However, there are several types of units with different uses and capabilities; certain units can't leave certain areas, and each works better in a different sort of terrain. Most rules concern relationships between players and their control of areas, but there are special rules for Irish Gallоglass, the English Army, certain leaders (such as William Wallace), changes of allegiance, and so on.

COMMENTS: An interesting game on an esoteric subject, *Robert the Bruce* has quite a bit of color. The rule book contains a good deal of history on both the events and the participants. The rules are short, but they have some holes, especially in the terrain department. The map, while nicely done, is difficult to read. Although the two games differ substantially, *Robert the Bruce* has much of the feel of *Kingmaker*. The former is less successful, largely because the actions of the players are more restricted. Even so, the situation is interesting, and the strategic possibilities are manifold. Involved players can enjoy an entertaining evening with it.

EVALUATION:
- Presentation—Very Good
- Rules—Fair
- Playability—Very Good
- Realism—Good
- Complexity—5

OVERALL EVALUATION: Good

The Siege of Constantinople (1978)

PUBLISHER: Simulations Publications, Inc.

SUGGESTED RETAIL PRICE: $10 (plastic box)

SUBJECT: This game of tactical operations in the Ottoman siege of Constantinople includes only land operations and is not really complete without the naval variant that was published in *MOVES* magazine #37.

PLAYING TIME: As long as you can stand it; otherwise one to two hours.

SCALE: This is a tactical-level game with operational overtones. Counters represent a few hundred men (very roughly), turns two days, and hexes two hundred yards.

SIZE: The 32″ x 21″ map uses 19mm (¾″) hexes.

BALANCE: This depends on who gets the lucky die rolls.

KEY FEATURES: The victory conditions rest on whether the Ottoman player can get twenty-five Attack Points inside the city. To reach this conclusion, players must consider the effects of sappers, siege ladders, tunnels, catapults, and cannon—all of which are covered by die rolls against various tables.

COMMENTS: This is one of the most disappointing games of recent memory. It was intended as an "issue" game in *Strategy & Tactics* magazine to raise interest in SPI's new (and, as it happens, far better) series, *The Art of Siege Warfare*. When the design and development fell well behind schedule, the naval module, which was to re-create the Turkish assaults on the harbor, was discarded to speed the game toward its deadline. More problems arose, further changes were made, and the focus of the game went in and out like a zoom lens. In the end, it was published in *Strategy & Tactics* #66 to resounding disapproval. (It placed third in a poll conducted by *Fire & Movement* magazine for "Turkey of the Year.") The game map distorts the scale and location of features of the city wall, the only really important piece of terrain in the city. This was done to allow a map of the harbor to join it, but it's meaningless color without the naval module. The game proceeds as a die-roller until the Ottomans breach the wall. Then things get interesting for a couple of turns, and the Ottomans win. Or they *don't* breach the wall, and the Ottomans lose. That's it. It would be quicker to flip a coin.

EVALUATION:
Presentation—Good
Rules—Fair
Playability—Fair
Realism—Good
Complexity—6

OVERALL EVALUATION: Poor

Timbuktu (1978)

PUBLISHER: Imperium Publishing Co.

SUGGESTED RETAIL PRICE: $9 (resealable plastic bag)

SUBJECT: This is a multiplayer game of medieval African politics, covering events in the West African Sudan, A.D. 1050–1500.

PLAYING TIME: A complete game takes about ten hours, but there are shorter scenarios.

SCALE: This is a highly strategic-level game using tribes and peoples. Each turn represents fifty years.

SIZE: The 22″ x 28″ game map uses areas rather than hexes.

BALANCE: This aspect is good.

KEY FEATURES: The game, which covers a unique subject and one that seems to garner little interest in the hobby, uses time-honored systems and mechanics to simulate its era. The combat system uses standard infantry and cavalry units and a simple Attacker Eliminated/Defender Eliminated type of Combat Results Table. Rules cover controlling provinces, revolts, collecting revenue, and maintaining empires. There are random events, and players may seek "Glory Points." The sequence of play is quite long and includes an "Administrative Segment" for handling all the paperwork and money problems.

COMMENTS: This is a nice and not overly complex power-politics game on an arcane but colorful subject. Much of the game's intricacy is left to the players, who, as is usual in games of this sort, make their own problems. Despite the highly strategic nature of the game, which tends to abstract many of the historical details, the players can gain a fair insight into a neglected part of history. To aficionados of this genre, the familiarity of some of the mechanics will be more of an advantage than a drawback. If the game is nothing spectacular, it is surely a solid, playable effort.

EVALUATION:
Presentation—Good
Rules—Good
Playability—Very Good
Realism—Fair
Complexity—6

OVERALL EVALUATION: Good

Viking (*1975*—a redesigned version of *1971's Dark Ages*)

PUBLISHER: Simulations Publications, Inc.

SUGGESTED RETAIL PRICE: $12 (plastic box)

SUBJECT: As part of the PRESTAGS system, *Viking* covers tactical warfare in the Dark Ages, A.D. 700–1300. As scenarios, it includes the battles of Tours, Clontarf, Constantinople, Stamford Bridge, Hastings, Manzikert, Durazzo, Dorylaeum, and Liegnitz.

COMMENTS: For information on this game, see the evaluation of *Chariot*, which covers the entire PRESTAGS system. *Viking*, however, is an exceptional selection because most of the battles are very exciting.

William The Conqueror (1976)

PUBLISHER: TSR Games, Inc.

SUGGESTED RETAIL PRICE: $11.98 (boxed)

SUBJECT: The land battles of Stamford Bridge and Hastings, A.D. 1066.

PLAYING TIME: Each battle takes several hours to complete.

SCALE: This is a tactical-level battle game using groups of men (there being no particular organization in those days). Turn length is an hour. Each hex measures about one hundred yards.

SIZE: The 22″ x 27″ vinyl game map uses 25mm (1″) hexes.

BALANCE: Fair. Harold has the upper hand at Stamford, and William can usually win at Hastings—but not by that much in either scenario.

KEY FEATURES: This is an unusual game—and that is its charm and its downfall. The counters are round—a nice touch—and quite large. The simple rules cover movement and a unique combat system; it uses something of a slide rule that combines strength with terrain to compute the results. (It is not as complex as it sounds.) This system eliminates dice and, to a great extent, chance. Leaders play an important part, and on a minor level there are rules for morale as well as the usual facing, fire, and melee procedures.

COMMENTS: *William the Conqueror* is another of TSR's attempts to break into the "historical" game market. As a production, it is admirable; as a game, it leaves a lot to be desired. The idea of nonrandomized combat may seem to be a good one, but in practice it's a big yawn. Much of battle *is* fairly random—especially in the Middle Ages, when commanders had little control over their troops while the battle raged. Randomness is also a prime enjoyment factor; to be able to predict all outcomes is to remove the element of surprise and much of the tension from a game. In addition to these problems, historicity has been bent a bit—albeit not enough to annoy those not intimate with the two battles. The map of Hastings is not particularly accurate, and the initial deployments are open to question. (But so are the ones suggested by historians.) Disappointing.

EVALUATION:
Presentation—Very Good
Rules—Good
Playability—Very Good
Realism—Fair
Complexity—5

OVERALL EVALUATION—Fair

Yeoman (1975—a redesigned version of 1970's *Renaissance of Infantry*)

PUBLISHER: Simulations Publications, Inc.

SUGGESTED RETAIL PRICE: $12 (plastic box)

SUBJECT: *Yeoman*, part of the PRESTAGS system, covers tactical warfare in the Renaissance Era, A.D. 1250–1500. Battles covered in this game include Bannockburn, Crécy, Barnet, Agincourt, and others.

COMMENTS: For information on this game and on the entire PRESTAGS system, see the evaluation of *Chariot*. Yeoman is a bit more complex than the others in the system because combat became more complicated with the introduction of gunpowder.

CHAPTER 8

Battles Before Bonaparte

WHEN THE sixteenth century dawned, only Austria—which, coincidentally, was the seat of the Holy Roman Emperors—was in any way a "stable" nation, with constant borders and an ordered succession to the throne. The rest of middle Europe was a patchwork quilt of principalities, duchies, and electorates that were all vassals of the Emperor—technically. Political intrigue was a fuse burning slowly toward an explosion that would scorch all of the continent in the next century.

Other areas of Europe were no more stable. Only Britain, France, and Spain had even territorial security, and these three were the major rivals for power and each other's territories. France and Spain contested portions of the Mediterranean trading empire in Italy. England and France argued and fought over continental territories claimed by the English throne. England and Spain quarreled over affairs in the Netherlands, which was claimed and ruled by Spain. And these were the "stable" countries!

The Spanish Army had been forged in the battles that finally expelled the Arabs from Granada. Though the soldiers who fought the Moors were gone from service by the second decade of the century, the systems of command and training introduced by the military nobles were preserved. New armies used in the Spanish Mediterranean adventures and in the Americas meant a steady flow of veteran soldiers—really the first professional troops seen in Europe since the Romans. The Spanish forces were nearly unstoppable. They were also very inventive.

The first Spanish armies were built around square formations of pikemen and swordsmen supported by wings of lightly armored cavalry carrying sabers or lances. The tactics favored would force the enemy to break his army against the Spanish squares, which were all but impenetrable. The cavalry was not wasted against opposing pikemen. Instead, it was used to harass the enemy, engage reserves—especially fresh cavalry—and pursue the infantry once it had broken ranks. These were all lessons learned by hard experience with the Moors, who had used the same tricks against the Crusaders. When small firearms—the arquebus, musket, and rifle—began to be used, the Spanish modified their formation. They placed gunners at the four corners of the square, where they had an open field of fire at advancing troops from any flank. With experience, the formation came to contain gunners who could load and fire from within the square in alternate volleys.

The *tercio,* as the formation was called, became the basic organizational unit of Spanish armies. The protection that it offered to gunners gave it battlefield superiority. Although early firearms didn't have great accuracy or range, the human wall of pikes and swords allowed Spanish gunners to exercise their trade against targets at conveniently close range. Because other European nations were slow to catch on, the Spanish sphere of influence in Europe grew.

The only surviving game that covers even a portion of this early era is Simulations Publications' *Yeoman.* Although the real focus of *Yeoman* is the battles of the fourteenth century, the lessons that were learned therefrom had applicability to the following two hundred years. A better game for the sixteenth-century period was its predecessor, *Renaissance of Infantry.* Using the same basic design, it demonstrated the superiority of the Spanish *tercio* against the French and Neapolitans.

Musket & Pike shows how the Spanish square was replaced by newer tactics. Although somewhat slow in play, *Musket & Pike* deserves attention. Its scenarios based on the Thirty Years' War, however, have largely been displaced by later games covering that conflict.

Meanwhile, back in the Western Hemisphere, the Spanish had begun an empire that was to make them wealthy. The gold and silver of the Incas and Aztecs was funneling into the treasury. Tobacco, rum, and some corn were coming back from the new lands. Native crafts and the gaudy plants of the Amazon Basin were being spilled into the cities in a confusion of colors. All of this was waiting in the Americas. Sometimes there might be the slight nuisance of killing the owners, but this was simple for Europeans. Their diseases ravaged the Indian populations, and their gunpowder and horses did the rest. With fewer than two hundred men, Francisco Pizarro plundered the Inca Empire of its riches. The same was done in Mexico, Brazil, and Argentina. The real riches of the American continents were unsuspected, but the legends of cities paved with gold kept the adventurers coming.

The New World provided more warfare than anything else in the era. Although in the first centuries only a few armed conflicts occurred in North America—and those minor—battles over colonial possessions in South America and the Caribbean were as commonplace as piracy. The riches of the hemisphere funded European wars well into the seventeenth century, and in the eighteenth century wars were fought for the continents themselves.

Games of exploration are uncommon, and only one exists to model the European expansion into the Americas. This is *Conquistador!* The game has interesting provisions for the competition and interaction of the French, English, Spanish, and Portuguese, but it shows very little real exploration. The major occupation of the players is to exploit the riches, most of which are marked on the map sheet. Between the days of Columbus and Napoleon were three centuries of power politics, with many nations of Europe jockeying for position in a world that had increased in size and wealth by more than one hundred percent. It's a shame no game designer has yet treated this global struggle.

Another conflict was brewing in Europe at this time, too. It began on October 31, 1517, when Martin Luther nailed ninety-five theses of religious philosophy to the door of the church at Wittenberg. *A Mighty Fortress*, published by Simulations Publications, is a good, if obscure, game of the social aspects of the Reformation.

In 1618, the nobles of Bohemia deposed their king, Ferdinand, who was strongly anti-Protestant, and gave the throne to Frederick, whose reign was to be so brief it earned him the sobriquet, "The Winter King." Ferdinand, who also happened to be the Holy Roman Emperor, decisively defeated the armies of Frederick within a few months. The war over the Bohemian succession was not so quickly settled, though. It grew and grew, finding new fuel when old issues burned out. Known as the Thirty Years' War, it was a collection of small wars involving virtually every nation on the continent.

The Thirty Years' War was the first "modern" war and the first heavily influenced by merchants and trade, rather than being directed at the simple occupation of trade centers or routes. It fixed the idea of having a professional army in the minds of kings, and it established the importance of the rifle and the cannon as weapons. The conflict not only decisively settled the issues of the Reformation and Counter-Reformation; it also shifted the teetering European ascendancy away from Spain and toward France and England.

SPI's *Thirty Years' War* QuadriGame and the "extra" *Breitenfeld* are quite good simulations of the warfare. They show the special problems of the conflict in some detail. Unfortunately, they don't show the tactical changes that displaced the Spanish square as the standard of organization and deployment. Such niceties have been refined into the combat factors of the units.

After the Spanish system fell into disrepute and disuse, the Swedish system was widely adopted. It emphasized a detailed chain of organization and command within regiments, which were the basic units. It was practical, and it developed a more fluid style of combat that was better suited to the rapid movements of troops and cannon. Its ascendancy was brief, however. It fell by the wayside thanks to Frederick the Great.

Frederick was the most successful commander of pre-Napoleonic modern times. His success, though, came less from innate skill as a commander or tactician than from two other abilities. First, he was able to recognize and develop someone else's good ideas. He learned far more quickly than his opponents, and his evolving styles were a mystery to most of them. Second, he was a wise organizer, and he was able to develop the military machine left him by his father and grandfather. The most important thing he did was to establish strict military discipline. This, with extensive and intensive drill, made Prussian troops capable of rapidly changing formations and responding to the commands of officers, who were also well trained and could spot opportunities in the midst of battle. No other army of its day could maneuver as quickly and skillfully as Frederick's Prussians.

The grand-tactical considerations of Frederick's campaigns are covered by SPI's *Frederick the Great*, but the depth of strategic subtlety reached before the nineteenth century was not much deeper than that achieved by Wallenstein or Gustavus Adolphus in the Thirty Years' War. This means that gamers are probably much more capable than the generals they emulate in games like *Frederick the Great*.

On the other hand, tactics of that time

were less developed than the peak reached by Napoleon, so a similar complaint can be raised for smaller-scale games. SPI's *Grenadier* covers not only the eighteenth and late seventeenth centuries but the Napoleonic period as well. It emphasizes the importance of deployment from column to line, the speed at which artillery could be limbered or unlimbered, and the importance of breaking an advancing line into skirmishers at the proper moment. The picture it presents has little to do with the tactics of Napoleon, and, more to the point here, an average gamer can probably maneuver better than the game requires.

Warfare in North America was different from the European model. The Indians and the American colonists were both necessarily disposed to "ad hoc" applications of tactics. The American revolutionaries were indebted to the Indians for more than corn; the Indian tactics that had been effective against the colonists were devastating against the British. The redcoats and Hessian "mercenaries" were unprepared for anything other than the meeting engagements pioneered by Frederick of Prussia. Guerilla warfare and its ramifications were unknown to them.

A great number of games based on the American Revolution exist, many fairly good. Most, however, are overly indebted to Napoleonic design techniques or to undue abstraction from an urge to escape them. The most outstanding games of warfare on this period are *1776* and *Quebec, 1759*. Both emphasize playability, from which their reputations arise. Tactical games of the Revolution are few, and good treatments of the constraints of battle are even rarer.

Lille is too new a game to have attracted much response. It is an abstract design of seige warfare that plays surprisingly well. Despite the complications evident in the rules, there is a simple pattern to play. This game shows the last gasp of feudalism in the destruction of a truly mighty fortress, the masterwork of Sébastien, Marquis de Vauban, the greatest military engineer of the seventeenth century.

There have been many attempts at sea games of the era, of which Avalon Hill's *Wooden Ships & Iron Men* and SPI's *Frigate* are among the best known. Based on the same frame of design as all the other SPI naval games, *Frigate* concentrates on generally earlier battles than *Wooden Ships & Iron Men*. *Frigate* also differs in scale and style of play. Lacking in the games is a design featuring the evolution of tactics from the oared ships of the millennium through the sailing vessels of the seventeenth century.

The first two centuries of the modern era were dominated by experimentation in warfare. New tactics were tried and discarded as the military learned to cope with the social and industrial revolutions that were transforming peace and war alike. Games can only capture the spirit of the times in individual treatments. There will never be a "system" for seventeenth-century tactics, though systems for the Thirty Years' War or the English Civil Wars may become established. The nature of the era was change, and there were too many distinctions for one game system to do it justice. With so much there to game, it's too bad Napoleonics are so much more popular.

Evaluations

Breitenfeld (1976)

PUBLISHER: Simulations Publications, Inc.

SUGGESTED RETAIL PRICE: $4 (resealable plastic bag)

SUBJECT: *Breitenfeld* simulates the first battle between the army of the imperialists and the Swedish Army of Gustavus Adolphus. It occurred north of Leipzig on September 17, 1631.

PLAYING TIME: The game takes one to three hours.

SCALE: This is a battle-level (grand tactical) simulation. Cavalry and infantry units represent regiments and brigades, while artillery is grouped into loose batteries. There are 175 meters per hex and forty-five minutes per turn.

SIZE: The 17″ x 22″ map uses 19mm (¾″) hexes.

BALANCE: This is good, although there is a slight bias in favor of the Swedish forces.

KEY FEATURES: *Breitenfeld* uses a simple system with the standard move/attack turn sequence to allow the game to be played to conclusion in a short period of time. Strict limitations are placed on each side for the first four or five turns, when only artillery fire is allowed, although certain events may release one or both sides from the movement restrictions. Zones of control don't affect movement, but combat between adjacent opposing units is mandatory. Special rules cover battlefield visibility and Swedish cavalry charges. Morale is affected by cumulative losses.

COMMENTS: More and more games which require a great investment of time to play to conclusion are being published. *Breitenfeld* thus serves a need too often neglected in wargaming. While its short playing time is certainly due in part to the folio format in which it is presented, it was clearly the designer's intent, as well. *Breitenfeld* utilizes the same system employed in the *Thirty Years' War* Quadri-Game—itself a variant of the *Napoleon at Waterloo* game system. It's a simple, fast-moving game suited for the novice or an experienced gamer looking more for fun than for a challenge.

EVALUATION:
- Presentation—Good
- Rules—Very Good
- Playability—Very Good
- Realism—Good
- Complexity—4

OVERALL EVALUATION: Good

Conquistador! (1976)

PUBLISHER: Simulations Publications, Inc.

SUGGESTED RETAIL PRICE: $10 (plastic box)

SUBJECT: The exploration of the Americas, 1495–1600.

PLAYING TIME: The game can take from two to twenty hours.

SCALE: This is a strategic game with one-hundred-mile hexes and five-year turns. Units vary greatly from individual explorers or missionaries to troop detachments or groups of one to five ships.

SIZE: The 22″ x 32″ map uses 16mm (⅝″) hexes.

BALANCE: In multiplayer games, the English and/or Spanish are favored; the French and Portuguese are always at a disadvantage.

KEY FEATURES: *Conquistador!* is not strictly a wargame. Instead, it's a competitive model of the early stage of modern exploration, from the time of Columbus until the time when the Thirty Years' War tied up the energies of Europe. Large elements of chance are built in to convey the uncertainties facing Europeans in the New World. Since the terrain features on the map are fixed and unmoving, the game is necessarily devoid of the searching the explorers really had to do. (Compare with *Source of the Nile.*) Therefore, the rewards accruing to a player who secures one of the objectives have been made uncertain. Furthermore, other tables reflect the changing political fortunes back home (for example, a new monarch may cut you off in the Americas). A workable solitaire version is also provided.

COMMENTS: This may be the best of Richard Berg's designs, although it's been largely ignored by the hard-core fans. The conflict it simulates is simpler and more subtle than they are used to. The techniques of victory are not the techniques of battlefield superiority. Of course, it wouldn't be a Berg game without plenty of random events, but in multiplayer games this element of chance rewards good planning. You must allow for changing fortunes both on and off the map. On the other hand, and this is a flaw of some significance, there is little reward in real "exploration." Players are sure to be too cautious to go chasing after the Seven Cities of Cibola unless forced by random events. There are more victory points in settlement and steady expansion of control than in the adventure of discovery. This, alas, removes some of the glamour the game might have carried; it's too cynically realistic. The game is an oddity, but it's also a lot of fun, and it contains a challenging, interesting, and rewarding solitaire system. What more can you ask? (. . . Well, maybe a slightly shorter game.)

EVALUATION:
- Presentation—Excellent
- Rules—Good
- Playability—Good
- Realism—Good
- Complexity—7

OVERALL EVALUATION: Good

Frederick the Great (1975)

PUBLISHER: Simulations Publications, Inc.

SUGGESTED RETAIL PRICE: $12 (plastic box)

SUBJECT: The game deals with the land campaigns of Frederick the Great of Prussia, 1756–1759.

PLAYING TIME: Each scenario takes six to eight hours.

SCALE: This is an operational game in which strength points equal about twenty-five hundred men. Each turn represents fifteen days, and each hex covers about fifteen miles.

SIZE: The 21″ x 33″ game map uses 16mm (⅝″) hexes.

BALANCE: The balance is excellent.

KEY FEATURES: The heart of *Frederick the Great* is its interwoven sequence of play, which allows the defender to force-march during the opposing players's turn. This establishes the game as one of maneuver and position. Combat is by odds/ratio with losses based on size of force involved. Leaders are all-important: you can't move without them. Although they are rated for ability, you must use higher-ranked leaders, regardless of their skills. Supply is quite restrictive and sets the tone for much of the game, as players try to establish depots so that they can gain at least a modicum of maneuverability. Along with some simple morale rules, sieges, garrisons, prisoners, the honors of war, and wintering are covered.

COMMENTS: Aside from its rather drab physical appearance—a result of its being done before color became the vogue—opinion is quite divided on *Frederick the Great*. It is arguably one of the finest simulations on the market. However, since the object of the game is to avoid combat, most people find six to eight hours of that sort of thing quite boring. One cannot really blame them, but in its way it has few peers. The game, which established designer Frank Davis' reputation, is not only a superb battle of wits, but it is also an excellent evocation of eighteenth-century warfare. The major elements of campaigning in the field—poor supplies, dumb commanders, and the knowledge that to win you must outmaneuver the opposition—are all present. While it is by no means to everyone's taste, those looking for enlightenment and its special kind of mind-boggling challenge need look no further.

EVALUATION:
Presentation—Fair
Rules—Good
Playability—Good
Realism—Very Good
Complexity—6

OVERALL EVALUATION: Good

Frigate (1974)

PUBLISHER: Simulations Publications, Inc.

SUGGESTED RETAIL PRICE: $10 (plastic box)

SUBJECT: The game focuses on naval combat from 1702 to 1812.

PLAYING TIME: A game can take thirty minutes to more than ten hours, depending on scenario complexity.

SCALE: *Frigate* is a tactical simulation with units representing individual ships. Each hex represents two hundred meters, and each turn spans approximately nine minutes.

SIZE: The game includes six 10″ x 10¾″ isomorphic map sections that form a 20″ x 32″ playing surface. A slightly larger playing area is desirable, as battles tend to drift downwind. Hexes are 16mm (⅝″) across.

BALANCE: This varies; balance in each of the twenty scenarios is described in the playtest notes in the game.

KEY FEATURES: The simultaneous movement system requires a written plot for movement *and* combat. Rules for movement in formation relieve the burden to some degree, but the playing time increases predictably with the number of ships in the scenario. Combat is very abstract: hits affect either gunnery strength or movement capability, and the effects are the same for all ships, regardless of size. Crew efficiency affects both the capacity for movement independent of formation and the gunnery rating, which determines the Combat Results Table used. Command control rules, which may require the random movement of affected ships, are tied to a ship's efficiency rating and its present hex number. Preservation levels limit the durability of each side's force as a whole and thus the maximum length of a scenario. Optional rules are sprinkled throughout the text. Designer's, developer's, and historical notes are also included.

COMMENTS: *Frigate* is a game of maneuver.

While maneuver is vitally important in any naval battle—especially when movement is largely dependent on the wind—many other factors that could add immeasurably to the game have been so abstracted as to render them devoid of much of their attraction. This is not to say that *Frigate* is not a good game, but that it does not fully realize the potential of the situations it presents. Minor shortcomings include the location of several important tables only within the body of the rules and the decrease in solitaire playability due to the written movement system. More annoying is that, because of the written combat plots, many turns are wasted firing into an empty sea. However, the fluidity of play in *Frigate* does give a real appreciation for the difficulties in maneuvering a force of several independent units while attempting to maintain a positional advantage. Most of the optional rules should be included, as they add interest. The beginning scenarios aid in learning the rules, but the intermediate scenarios are probably the best entertainment for the time invested.

EVALUATION:
Presentation—Good
Rules—Good
Playability—Good
Realism—Good
Complexity—6

OVERALL EVALUATION: Good

Guilford Courthouse (1978)

PUBLISHER: Game Designers' Workshop

SUGGESTED RETAIL PRICE: $5 (resealable plastic bag)

SUBJECT: The game focuses on a land battle in the American Revolution in March, 1781.

PLAYING TIME: Game time is two hours.

SCALE: This is a battle-level (grand-tactical) game utilizing companies. Turns are roughly thirty minutes, and hexes represent one-hundred-and-fifty to two-hundred yards.

SIZE: The 17″ x 22″ game map uses 19mm (3/4″) hexes.

BALANCE: Balance is difficult to determine, as it depends on the "plan" adopted by the player. A good British player should have little trouble whipping the poorly trained American force, but the former has difficult victory conditions to meet.

KEY FEATURES: Manpower levels determine both firepower and melee strengths. Considering the complexity level of the game, the combat system is remarkably sophisticated. Leaders are used, and the emphasis is on morale. There are special rules for the several types of units.

COMMENTS: This is in GDW's Series 120 games—meaning it should play in two hours. Thus, complexity is limited. Given this, the game is interestingly subtle and covers many areas of eighteenth-century warfare in nice detail. However, the system is better than the game, and the fact that there is no terrain key to decipher the map doesn't help. Moreover, both sides seem to have optimum strategies, and play can become stereotyped (and, tangentially, quite dissimilar to the actual battle). Despite its ease of play and excellent system, many players will probably lose interest quickly.

EVALUATION:
Presentation—Good
Rules—Very Good
Playability—Very Good
Realism—Very Good
Complexity—5

OVERALL EVALUATION: Fair to Good (the game is not equal to the sum of its parts).

Lille (1978)

PUBLISHER: Simulations Publications, Inc.

SUGGESTED RETAIL PRICE: $9 (resealable plastic bag; also available boxed, as part of *The Art of Siege Warfare*, $27)

SUBJECT: The setting is the Siege of Lille in 1708, pitting the Dutch and English against the French.

PLAYING TIME: The game takes about six hours.

SCALE: This is a tactical-level game using brigades and cavalry squadrons. Each turn is the equivalent of five days. The hexless map is scaled at 1″ = 100 yards.

SIZE: The 22″ x 34″ game map uses no hexes or areas. It is simply a map of Lille.

BALANCE: This is fairly good, although the besiegers seem to have a bit of an edge.

KEY FEATURES: With no hexes or areas, this is a breakthrough simulation for board gaming. A tangential result is that the game map is quite striking. The attackers may move anywhere within their siegeworks—double-sized counters representing parallels, communications trenches, and gun galleries—which are constructed during the besieger's turn. The defenders move from position to position within the fort in anticipation of attacks. Artillery fire is ranged (with a rangefinder), and, when you finally reach the walls, melee is based on the morale of the units involved. There are rules for attrition, French field intervention, sorties, and leaders, but most of the rules pertain to siegework construction and movement, as both systems are new to wargaming.

COMMENTS: SPI took a big chance with their nonhex system, but to a great degree it works. The game truly develops—and looks like—a bird's-eye view of a siege, with its parallels and trenches sprouting and spreading. Of course, there is a bit of a problem getting used to any new system but, once assimilated, *Lille* is not that complex because the combat and fire systems are simple. The main problem is that siegework is, in strategic terms, a bore. The defender sits around dodging a few shells and wondering what will happen, while the attacker works valiantly to get his trenches as far forward as he can. Everything is aimed for the final big assault, which comes several hours after you start. Thus, *Lille* requires a bit of stamina until you reach the good part.

EVALUATION:
Presentation—Very Good
Rules—Good
Playability—Good
Realism—Very Good
Complexity—7

OVERALL EVALUATION: Good

Lobositz (1978)

PUBLISHER: Game Designers' Workshop

SUGGESTED RETAIL PRICE: $5 (resealable plastic bag)

SUBJECT: The scene is the Battle of Lobositz in 1756, part of the Seven Years' War in Europe.

PLAYING TIME: It takes two to three hours to play.

SCALE: This is a grand-tactical simulation using battalions. The turn equivalent is about an hour, and the hexes represent about four hundred yards.

SIZE: The 17″ x 22″ map uses 19mm (¾″) hexes.

BALANCE: The balance is quite good, more because of the poorly devised victory conditions than the actual play (in terms of which the Prussians have an edge).

KEY FEATURES: As with most GDW games set in this era, the emphasis is on morale. The play system is a basic one—Move/Rally/Fire/Melee—with simple Combat Results Tables. Good artillery rules and some nice weather restrictions add flavor, but morale is king.

COMMENTS: Although the situation is something of a set piece, *Lobositz* is a good, clean game with a pleasant eighteenth-century flavor. The problem is that the Austrians can get crushed on the field and still win the game; the victory conditions are ludicrous. If you ignore that aspect, this can be quite an engrossing game.

EVALUATION:
Presentation—Good
Rules—Very Good
Playability—Very Good
Realism—Good
Complexity—4

OVERALL EVALUATION: Good

Mercenary (1978—a new edition of a game originally published in England in 1975)

PUBLISHER: Fantasy Games Unlimited

SUGGESTED RETAIL PRICE: $6 (resealable plastic bag)

SUBJECT: The game is based on European power politics on land and sea, 1494–1560.

PLAYING TIME: An evening is required.

SCALE: This is a strategic game. Four units of any type equal an army. Turns are equal to six years.

SIZE: The original map was 23" x 29". The new one is slightly smaller, but it is in full color. Both maps use areas, not hexes.

BALANCE: When all five major powers are being played, balance is good. With fewer players, things are a little out of whack (as with *Diplomacy*).

KEY FEATURES: This is an ambitious game. The designers not only put in the usual diplomacy mechanisms—alliances, backstabbing, and so on—but they also have a system that accounts for advances in military technology, if you can afford it. Of further interest is the Diplomatic Scale, on which the influence of the major countries on minor ones can be affected by bribes, sacking—even marriage! As the title of the game implies, there are mercenaries for hire. The differential Combat Results Table gives the usual retreats and eliminations.

COMMENTS: There are a lot of interesting mechanics here—systems that were good enough to be borrowed by other games—and an interesting era. Despite the improved physical quality of the new edition, however, the game does not equal the sum of its parts. The rules are hazy—and that hurts. A lot is left to the players' imagination, as the designer has made the standard mistake of assuming the gamers know what was in his mind. (*Dungeons & Dragons* is the ultimate example of this.) For those particularly interested in this age or this genre, this could be a worthwhile purchase. Otherwise, try something else.

EVALUATION:
Presentation—Good
Rules—Fair
Playability—Good
Realism—Good
Complexity—6

OVERALL EVALUATION: Fair

Musket & Pike (1973)
PUBLISHER: Simulations Publications, Inc.
SUGGESTED RETAIL PRICE: $12 (plastic box)
SUBJECT: The game involves tactical combat on land, 1550–1680.
PLAYING TIME: This varies according to scenario, but generally takes three to four hours.
SCALE: This is a tactical game. Each unit represents one hundred to one hundred twenty-five men. Turns represent five minutes, with each hex one hundred meters.
SIZE: The 22″ x 34″ map uses 16mm (5/8″) hexes.
BALANCE: This is generally good, although there are some one-sided scenarios. (As in *Frigate,* each is rated for balance.)
KEY FEATURES: *Musket & Pike* generally follows the same format as the PRESTAGS games (see *Chariot*), although it preceded those games by several years. In fact, the elegance and facility of *Musket & Pike* led SPI to redo its old tactical games to comform with this system: hence, PRESTAGS. There are major differences, however, as gunpowder is in use in *Musket & Pike.* The Melee and Fire Combat Results Tables are quite similar, the results in both being to disrupt or eliminate units. There are range effects for various types of weapons and optional rules for such items as cavalry *caracolla,* squares, dismounting, and so on. The scenarios—each rated for complexity—cover such battles as Coutras, Breitenfeld, Lützen, Dunbar, and Naseby.
COMMENTS: This is a good—if somewhat outmoded—game. It is easy to play, and most of the scenarios are fun. It has a bit more flavor than most SPI PRESTAGS games, although the scenarios are less interesting. While it is not exactly state-of-the-art, nothing has been done to take its place, and it's worth looking into.
EVALUATION:
Presentation—Fair
Rules—Very Good
Playability—Very Good
Realism—Very Good
Complexity—5
OVERALL EVALUATION: Good

Quebec, 1759 (1978—a new edition of a game originally published by Gamma Two Games, Ltd., in 1972)
PUBLISHER: Avalon Hill Game Co.
SUGGESTED RETAIL PRICE: $12 (boxed)
SUBJECT: The setting is an English attack on Quebec, by land and sea, in 1759.
PLAYING TIME: The game takes two hours.
SCALE: This is an "operational" game. Units and time are abstracted. There are no hexes. The scale of the map is 1″ = 1,000 yards.
SIZE: The 11″ x 32″ mapboard uses irregular zones rather than hexes.
BALANCE: This is excellent.
KEY FEATURES: This game has a unique system using attractive domino-like wooden counters instead of the usual cardboard. Movement is simultaneous and limited to moving units from only one area. Battle occurs when both players have units in the same zone and consists of a series of die rolls. These produce casualties that wear down the strength of opposing units, which is kept secret from the other player (something like the family game *Stratego*). Other rules cover naval movement, Indians, and supply.
COMMENTS: This is an exciting game with little insight into history other than seeing

approximately what happened. The system really works well here (arguably better than the sibling games *1812* and *Napoleon*), and play is fast and furious. *Quebec, 1759* has the advantage of being a maneuver game with some real, old-fashioned slugging. There's considerable bluff and bravado and a generally good time for all.

EVALUATION:
- Presentation—Excellent
- Rules—Good
- Playability—Excellent
- Realism—Fair to Poor
- Complexity—3

OVERALL EVALUATION: Very Good

1776 (1974)

PUBLISHER: Avalon Hill Game Co.

SUGGESTED RETAIL PRICE: $12 (boxed)

SUBJECT: The scene is the American Revolution, on land and sea, 1775–1781.

PLAYING TIME: The shorter scenarios can be played in an evening. The full Revolution will take about eighteen hours.

SCALE: This is a strategic/operational game in which the units involved can vary from small companies to full regiments. (Each strength point equals three hundred to five hundred men.) Each hex covers about twenty miles, and turns are monthly.

SIZE: The two 16″ x 22″ game boards use 16mm (5/8″) hexes. A fairly large table is needed to play the campaign game.

BALANCE: Balance is a weak point in this game, especially in the shorter scenarios. For example, Greene's southern campaign plays exactly opposite to the historical results. The campaign also tends to favor the British, unless hidden movement is used (which we recommend).

KEY FEATURES: Avalon Hill's *1776* is an unusual game, and its unique system has never really been repeated. The combat system seems straightforward at first (odds/ratio, with fairly simple results), but optional rules add a wide variety of adjustments, and the tactical card system adds an aura of "outguessmanship" reminiscent of Poe's "The Purloined Letter." Good, extensive supply rules use magazines and rolling supply units. Both artillery and Indians get special treatment, as do river and ocean movements. The terrain is quite varied (a bit too much, perhaps), and the reinforcement system is accurate and excellent. Through its use of quick maneuver tied to vulnerable supply lines, the whole system manages to impart a good feel for eighteenth-century campaigning.

COMMENTS: When this game first appeared, it was widely criticized and generally disregarded. However, unlike most other games, *1776* has improved with age. The secret is to play only the Campaign Game using most of the optional rules. If you do that, you get a slam-bang simulation of eighteenth-century guerilla warface, with both sides in a contest of wits. The British have an immense army but horrendous supply lines; the Americans, holding on through bitter winters, hope that the colonial militia can keep the redcoats off balance. The shorter scenarios simply don't do justice to the game; most are silly. The campaign, however, deserves more recognition than it has received. *A warning note:* Several rules from the first edition—notably, British fire superiority—have been changed in later editions; some of these changes have been for the worse.

1776, *a game based on the American Revolution.* (PHOTO: AVALON HILL GAME CO.)

Evaluation:
- Presentation—Very Good
- Rules—Good
- Playability—Good to Very Good
- Realism—Campaign: Very Good; Scenarios: Fair
- Complexity—6

Overall Evaluation: Good to Very Good

Thirty Years' War QuadriGame (1976—includes the folio games *Lützen, Nördlingen, Rocroi,* and *Freiburg*)

Publisher: Simulations Publications, Inc.

Suggested Retail Price: $14 (plastic box; $4 per individual folio game)

Subject: The game includes four land engagements in the Thirty Years' War—battles set in 1632, 1634, 1643, and 1644.

Playing Time: The game takes three to four hours.

Scale: This is a battle-level (grand-tactical) simulation. Each strength point is the equivalent of seventy-five to one hundred men, each turn forty-five minutes, and each hex 175 meters.

Size: The four 22″ x 17″ game maps (one for each battle) use 19mm (¾″) hexes.

Balance: Except for *Freiburg,* balance is generally good.

Key Features: This game attempted to go one step further than *Musket & Pike* on a somewhat larger scale. Although the general sequence of events is the same, the feel is thus somewhat different. Artillery fire is quite different and serves mostly to disrupt units. There is no ranged fire; all other combat is melee. Leaders and morale play a significant role, and the whole effect is an attempt to re-create the paradox of stolidity and free-for-all that characterized so many of these battles. Each game has its own optional and additional rules to cover the individual situations.

COMMENTS: Although *Thirty Years' War* is more colorful than *Musket & Pike*, the battles in the quad are, with one exception, mostly blah. The fact that a different person designed each game doesn't help. *Lützen* is nice, but *Freiburg* is a disaster. (It seems that designer Steve Patrick's *bête noir* in life is to get stuck designing dreadful situations like this; he also got albatrossed with *Remagen*.) The system itself is nice and easy, and the games play fairly quickly, but there's nothing special here apart from that.

EVALUATION:
Presentation—Good
Rules—Good
Playability—Very Good
Realism—Good
Complexity—4

OVERALL EVALUATION: Fair to Good

Torgau (1974)

PUBLISHER: Game Designers' Workshop

SUGGESTED RETAIL PRICE: $8.75 (resealable plastic bag)

SUBJECT: The setting is a battle between the Prussians and Austrians in 1760, part of the Seven Years' War.

PLAYING TIME: About twelve hours are needed for a complete game.

SCALE: This battle-level (grand-tactical) game uses battalions and regiments. Each hex is two hundred yards, and each turn covers fifteen minutes.

SIZE: The 21″ x 31″ game map uses 19mm (¾″) hexes.

BALANCE: Balance is a sore point because it takes a superior Prussian player to beat the Austrian. (This is rather ahistoric, since the Prussians won, sort of—which reveals either the greatness of Frederick, the Prussian king, or a flaw in the design.)

KEY FEATURES: One of the first attempts to use rules germane to miniatures, *Torgau* has much of the flavor of that type of wargaming. There are a variety of formations and some fairly intricate stacking rules. The system uses movement points to integrate fire combat, which takes place during the movement phase, with movement. Since there is also *defensive* fire during a player's movement, quite a bit of firing goes on. The Fire Combat Results Table is an odds/ratio (attacking strength compared to the mass of the defender modified by its covering terrain) and produces step losses. Melee is fought within the same hex; it's also odds/ratio but yields losses, routs, and disruptions. Units' morale ratings are used at various times, and there are special rules for such units as dragoons, lancers, sappers, and fusiliers. The varied Prussian entry times are also covered.

COMMENTS: *Torgau* was an important—if flawed—game. Its descendants include *La Bataille de la Moskowa, Terrible Swift Sword,* and other big tactical games. *Torgau* has a wealth of design ideas but some poor development. There is also the problem with balance: the Prussians have a hell of a time getting on the board, and the Austrians take advantage of it. Even with all these problems, *Torgau* has remained a fairly popular game among detail freaks, despite its age. (Yes, it's only about five years old, but games age rapidly in this hobby.) Despite its overworked fire system and opaque rules, *Torgau* is a tense, if tough, game.

EVALUATION:
Presentation—Good
Rules—Fair
Playability—Fair
Realism—Very Good
Complexity—8

OVERALL EVALUATION: Fair but significant

Wooden Ships & Iron Men (1975—a revised edition of a game originally published by Battleline Publications in 1974)

PUBLISHER: Avalon Hill Game Co.

SUGGESTED RETAIL PRICE: $12 (boxed)

SUBJECT: It involves naval combat from 1776 to 1814.

PLAYING TIME: The game takes one to five hours.

SCALE: This is a tactical simulation with double-length counters representing individual ships. Each hex is very roughly two hundred feet across, and each turn spans about one minute.

SIZE: The 22″ x 28″ mapboard uses 16mm (5/8″) hexes.

BALANCE: The game provides twenty-three scenarios of varying balance along with provisions for designing your own.

KEY FEATURES: *Wooden Ships & Iron Men* is the definitive simulation of naval warfare in the Age of Sail. Virtually all of the factors affecting combat at the ship-to-ship level are presented—all within a sequence of play that simplifies the decision-making process. Movement is by written plot, but the modest movement factors of ships under battle sail and the small number of ships involved in most scenarios make this far less an annoyance than one might expect. Ship's logs are a record of movement plots, gun loading, and miscellaneous notes concerning special damage received, anchoring, and so on. Gunnery combat considers target location (hull or rigging), range, field of fire, type of ammunition, crew quality, number of guns, critical hits, and collateral damage. Melee is performed between fouled or grappled ships. Optional rules detail limited communication between al-

Wooden Ships & Iron Men, *published by Avalon Hill.* (PHOTO: AVALON HILL GAME CO.)

lied commanders, wind changes, towing, guns obscured by falling rigging, exploding ships, taking soundings and running aground, barnacled hulls, and even the effects of scurvy on crew size! Designer's notes and a sample game will aid the newcomer in understanding the game.

COMMENTS: This game is a fine simulation. Two turns of this game speak volumes about the significance of wind direction for sailing ships-of-the-line, and period tactics are *de rigueur.* Purely for the feel of *being there,* this game is unsurpassed. With a small loss of playability, a variety of detailed rules increases the variables each commander must consider, as well as the options available to achieve his desired goals. The scale of the game is such that gunnery can have a telling effect even at long range. Melee is furious and bloody for all concerned, and the interaction of crew quality with actual crew size adds much to the illusion of reality. As usual, however, the necessity for written movement and melee plots reduces the solitaire playability, but for a fun-filled few hours and a feeling of real involvement *Wooden Ships & Iron Men* is tops for two players.

EVALUATION:

Presentation—Very Good
Rules—Very Good
Playability—Very Good
Realism—Excellent
Complexity—7

OVERALL EVALUATION: Very Good

CHAPTER 9

The Napoleonic Era

NAPOLEON BONAPARTE and the warfare of the Napoleonic period have fascinated wargamers since the hobby began because there were immense changes in the nature of warfare in both the tactical and strategic spheres. The French Revolution and Napoleon brought an end to the Baroque minuet that was war in the eighteenth century and ushered in the era of the nation at war. In addition to being an innovator in the practice of war, Napoleon was certainly one of the world's greatest military leaders. It is almost impossible to look dispassionately on the titanic struggle waged by his military genius and vigor against virtually all of Europe for two decades.

Although Napoleonic warfare stands at the end of a period of chivalry and color on the battlefield, it shares the romance of warfare with earlier periods. Picture the charge of a gaily clad regiment of hussars, swords raised high, sweeping across the battlefield to the sound of trumpets. As inaccurate or incomplete as the romance may be, it fires the imagination. Taken together with the great Emperor, it creates an irresistible magnetism most wargamers cannot ignore.

In the hobby of miniatures, the Napoleonic era overshadows all other historical periods. Thousands upon thousands of small lead figures have been dressed in the colorful uniforms of the armies of France and her enemies. Countless sets of rules for tabletop battles have been prepared and used to fight engagements based on the period. In terms of player satisfaction and period flavor, these rules have been quite successful. As we have pointed out, however, there are distinct limits to what you can do with miniatures. The typical miniatures battle deals with basic tactical units. Above a certain size a battle takes up too much room, too many troops—and too much money.

Board games set in the Napoleonic era have been most successful at the other end of the scale: the very large battle. Here the complexities of warfare in this period tend to fall away, and the flavor of battle can be gained from a very simple simulation, such as Simulations Publications' *Napoleon at War* QuadriGame. The result is a highly playable game that still is able to illustrate the movement and battlefield strategy of the times.

Just as miniatures encounters tend to fail as they increase in scope, board games tend to fail when they attempt to simulate basic tactical actions; they simply don't capture the flavor of the period. SPI's *Grenadier* is an example of this. The results in *Grenadier* may tend to come out the way they did in real life, but what is done with the forces provided in the game bears no relationship to Napoleonic-period tactics. Almost inevitably, later board games approaching this lower end of their effective range have tended to resemble sets of miniatures rules condensed and transferred onto boards and pieces of cardboard.

THE WAR IN A NUTSHELL

If we were to characterize warfare in the Napoleonic period with a single word, it would be *maneuver*. Campaigns consisted of long marches culminating in brief and bloody battles, a few of which would decide the issue. The trick was to concentrate as many troops as possible at the critical point. Often an army would field many more troops in the theater but fail to achieve superiority on the battlefield. Napoleon's campaigns in Italy (1796–1797) are probably the best examples of this. Although constantly outnumbered by the Austrians, Napoleon was able to achieve parity or superiority of force on the decisive battlefields.

The same salient characteristic—maneuver—extended to the battle as well as to the campaign. The effective ranges of weapons were so limited that men didn't fall from hostile action until the final moment of movement. Although sometimes employed for bombardments lasting hours at ranges up to a thousand yards, artillery was most effective when wheeled up to canister range, a scant few hundred yards.

Napoleonic wargaming is so specialized and so dependent on the nature of the armies and tactics of the period that we should discuss these matters in some detail.

ARMS AND THE MEN

The Napoleonic wars saw millions of men under arms. As many as a quarter of a million might be concentrated for a major battle. Such large numbers required a considerable degree of organization to accomplish anything at all.

The infantry formed the bulk of any army and did most of the fighting. The basic tactical unit was the battalion. In the French Army toward the end of the era, it consisted of six companies totaling about nine hundred men, although casualties might have reduced this number considerably. Two companies were *elite* and were sometimes formed into special grenadier and *voltigeur* battalions. Grenadiers might be used to make special assaults, defend a critical point, or form a special reserve, particularly when the Imperial Guard was not on the field. Whether formed into special units or kept with the parent battalion, the *voltigeurs* would generally be employed as skirmishers.

Although dating back to the mid-eighteenth century in concept, the extensive use of skirmishers was an innovation of the French Revolution. While conventional armies depended on iron discipline and rigid formations, the zealous revolutionary armies sent forward loose lines of individual infantrymen who could better aim their fire, while presenting less of a target than an enemy drawn up into tight lines. Eventually each battalion had a company of *voltigeurs*, and whole regiments were designated *leg-*

ere: light infantry that could be used as skirmishers.

Most of the infantry and sometimes even the light battalions would, however, be found in tighter formations: line, column, or square. Troops in these formations could be more readily controlled than could skirmishers; they had more mass in an assault or when defending against an assault; and they were far less vulnerable to attack from cavalry.

The square was the formation of choice against a cavalry attack. It was literally square, with about four lines of infantry along each face of the formation. Well trained infantry in square formation could generally hold off any number of cavalry indefinitely.

The line, on the other hand, was designed to maximize firepower. The entire battalion would be arranged in two or three tightly packed lines firing together or, by sections, in an almost continuous rolling volley. Because they were armed with fairly inaccurate muzzle-loading muskets, the infantry depended on mass fire—a veritable wall of lead—to be effective. At close range against a similarly packed formation, such a volley could cause casualties as high as fifty percent because some of the heavy lead balls passed through more than one body.

The disadvantage of the square and the line formations was the difficulty in maneuvering them, particularly across anything more than a parade ground. The solution was the column. This was not a column of route, as is sometimes imagined, but a handy set of lines (somewhat looser than "line" formation), as shown in Fig. 9–1. In the French Army, the column was generally nine men deep and about ninety men wide. Besides being more maneuverable, it allowed an attacker to mass more men at the critical point than a line formation would.

Although some light troops were equipped with rifles, the primary infantry weapon was the musket. Since this weapon took a longer time to reload, a bayonet was always attached. Despite this, very few soldiers died of bayonet wounds; the bayonet's effect was mainly psychological. While good troops with clean muskets could fire up to five shots per minute, one round a minute was more typical.

Cavalry might constitute ten to twenty percent of an army. During the march, the cavalry's principal duties were reconnaissance and screening the army from enemy patrols. In battle, its role was shock action: the mounted charge. This was generally ineffective against an infantry square. Against the flank or rear of a unit in line or column, however, it could be devastating. In the Spanish campaigns, two squadrons of cavalry broke an entire division of Spanish infantry when they charged the rear of the unit.

Fig. 9–1
French Battalion in Column

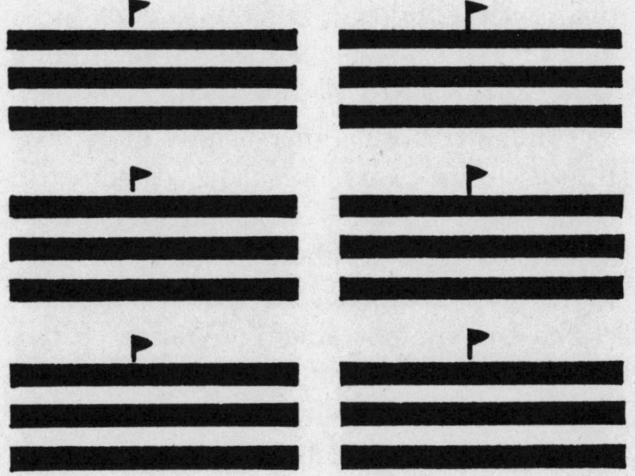

Cavalry was generally divided into heavy (big men on big horses) and light (smaller men on smaller horses). Heavy cavalry was used almost exclusively as battle cavalry; light cavalry was used more for reconnaissance. Members of both were armed with a saber and, in some cases, a lance. Some heavy cavalry men (*cuirassiers*) were equipped with body armor—a helmet and breastplate or back-and-breast—that was sufficiently thick to stop a musketball fired at medium range. Other cavalry, primarily dragoons and lights, also carried a musket and sometimes one or two pistols for use when dismounted or while on picket duty.

The basic tactical unit for cavalry was the squadron, which was typically understrength and closer to a hundred twenty men than the nominal three hundred. In addition to being part of combined arms formations, French heavy cavalry was normally segregated into a cavalry reserve to be saved for times when it could be employed en masse.

Artillery provided long-range firepower and caused a large part of battlefield casualties. Field artillery came in sizes of four to twelve pounds, according to weight of shot, and was grouped into companies of six to twelve guns. In battle, these guns might be gathered into large batteries of forty or more guns to pour a devastating barrage into hapless infantry formations before them. Artillery units were attached directly to combined arms units. Horse artillery, with somewhat lighter guns and mounted gunners, was available for use with cavalry.

The maximum effective range for a twelve-pounder firing round shot—a solid iron ball—was about a thousand yards. Grape shot and canister—multiple smaller balls fired with an effect somewhat like a shotgun blast—were much more effective against infantry but had a shorter range.

At the highest level, the French Army was divided into *corps d'armée,* each a small army consisting of two to four divisions of infantry, a brigade or so of cavalry, and five to ten companies of artillery. The corps was a natural and necessary formation for the large armies of the period. By 1815, every major army in Europe had adopted it.

TACTICS

A Napoleonic campaign would begin with each of the corps, perhaps half a dozen, moving to the general vicinity of the enemy. Some might be detailed to guard flank approaches; others might feint in a direction away from the main line of movement. Several corps, including the Imperial Guard, would advance along a central route, remaining close to the Emperor. When it became clear what the enemy would do, Napoleon would order his corps to concentrate where he wanted the battle to occur. An outlying corps might be detailed to engage a part of the enemy and keep it from joining up with the main body.

Once a major portion of the French Army had concentrated, it would begin a series of attacks on different portions of the enemy line, probing for weak spots and forcing the enemy to commit his reserves. Eventually Napoleon would launch the decisive assault to break the enemy line, "roll it up," and defeat each part in detail. Although this seldom resulted in the total destruction of the opposing force, the enemy army—battered by constant attacks, its reserves used up—would generally become demoralized and quit the field. Only a commander with the tenacity of Kutuzov or Blücher could salvage such a situation.

The mechanics of the attack are also worth investigating in some detail. At its best, the attack was a well coordinated affair combining all three arms. At its worst, it was simply a frontal assault by infantry or cavalry that was unsupported, across poor ground, and without even the advantage of superiority of numbers. In the latter case, it generally failed. When well done, however, the enemy would have been harassed for some time by French skirmishers. Cavalry would then attack—or threaten to attack—to force the enemy to form a square, pinning the army where it stood. Artillery batteries could set up within canister range and rapidly thin the ranks of the opposing infantry squares. The threat of cavalry attack would keep the enemy away from the guns. As the squares began to disintegrate under the withering fire, cavalry or infantry assaults would break through. French infantry would move into the vacated positions to hold the ground while fresh cavalry would pour through to spread panic in the rear.

Why didn't this technique always work? For one thing, as the numbers of men involved increased, French commanders resorted more often to main force and less often to skill; some of them never understood the techniques. As the quality of French troops declined, complicated attacks became less feasible. The necessary resources were not always there. For example, at Waterloo there was no cavalry left to support the final attack of the Imperial Guard. Finally, the enemies of France learned how to counter French techniques, most skillfully in the case of the British Army under the Duke of Wellington.

British light companies and light battalions neutralized the French skirmishers and kept them away from the main British line. Whenever possible, that line was deployed on the reverse slopes of hills to make it difficult for the French artillery to bring the squares of British troops under fire. The defense was active rather than passive. The British cavalry counterattacked the French cavalry whenever possible. Unengaged units moved forward and attacked advancing French columns, disorganizing them before they ever came near their objectives. Last but not least, the British infantryman was not impressed by the threat of the French bayonet. Confident in his training and his leaders, the British soldier was not so easy a target psychologically as a green Silesian *Landwehrman*.

THE WAR AT SEA

Napoleon was a general, not an admiral. He neither understood nor appreciated the significance of navies except as a diversion for his efforts on land. Perhaps, too, he realized that his naval forces had little chance against a British Navy that was much more advanced. At any rate, the Napoleonic era is not known for naval warfare. That the war at sea during this period can be made into excellent entertainment, however, is shown by *Wooden Ships & Iron Men*, which was reviewed in the previous chapter.

Evaluations

Austerlitz (1973)

PUBLISHER: Simulations Publications, Inc.

SUGGESTED RETAIL PRICE: $12 (plastic box)

SUBJECT: The Battle of Austerlitz, December 2, 1805.

PLAYING TIME: The game takes two to three hours.

SCALE: This is an operational-level simulation. The basic unit is the brigade. Hexes are four hundred meters across, and a turn represents one hour.

SIZE: The 22″ x 28″ board uses 16mm (5/8″) hexes.

BALANCE: Very good.

KEY FEATURES: *Austerlitz* also uses the *Napoleon at Waterloo* system. Artillery can fire from two hexes away, and only cavalry may withdraw from enemy zones of control. French units may be stacked three high, while Austrian and Russian units may be stacked to a total of two combat factors in a hex. In general, the Allied units are larger, less flexible, and slower than the French units. The mapboard is quite plain, although there are a number of villages, a castle, and an abbey that triple or quadruple defensive strength. Victory points are gained by killing enemy units or, in the Allied case, by exiting from one end of the board or the other. Morale is important (although the Allied player may well end up becoming demoralized even when he wins).

COMMENTS: The popularity of *Austerlitz* with the SPI staff (if not the gaming public) is justified; it's probably the best of the company's Napoleonic series of games and one of the most interesting around. Play is quick and clean; there are enough units to minimize the effects of chance, but few enough to avoid drudgery of piece-pushing. The best part of the game, however, lies in the strategic options it presents to both players. The Allied player can try for a big—but risky—win by heading off the west edge of the board, a marginal victory by retreating off the east edge, or he can try to beat the French on the field. The French player enjoys superior mobility and exterior lines, but he must make tricky decisions in allocating forces among his flanks and center. Both players must keep reserves and their options open for as long as possible. The game is usually in doubt until the end, offers a variety of outcomes, and presents innumerable opportunities for sharp tactical and strategic play.

EVALUATION:

Presentation—Good

Rules—Good

Playability—Excellent

Realism—Good

Complexity—3

OVERALL EVALUATION: Excellent

Borodino (1972)

PUBLISHER: Simulations Publications, Inc.

SUGGESTED RETAIL PRICE: $10 (plastic box)

SUBJECT: The Battle of Borodino between the French forces of Napoleon and the Russian forces commanded by Kutuzov, which took place west of Moscow on September 5–7, 1812.

PLAYING TIME: Two to three hours are

needed for the one-day scenarios; four to six hours for the campaign game.

SCALE: This is an operational-level simulation using brigades, divisions, and corps. Each hex equals four hundred meters, and each turn represents one hour.

SIZE: The 22″ x 34″ game uses 16mm (⅝″) hexes.

BALANCE: The first-day scenario favors the French, while the later scenarios favor the Russians. The campaign game is well balanced.

KEY FEATURES: *Borodino* uses the *Napoleon at Waterloo* system virtually intact but adds some new terrain features (rivers, streams, bridges), Russian redoubts, and special-movement rules for night turns. Optional rules cover the Imperial Guard, Moscow militia, and Russian reinforcements.

COMMENTS: *Borodino* is a very simple game based on a very complex battle. The commanders must exercise sound tactical judgment. The Combat Results Table is fairly bloody, and battlefield attrition is high. This tends to favor the Russians, since time is in their favor. The Imperial Guard release rule should be used to give the French an even chance in the campaign game. The Moscow militia adds considerable strength to the Russians and, if included, can tip the balance in their favor. Although the game is a far cry from the current state of the art, its challenges and the simple system used make it an entertaining and fast-moving game for novice and veteran gamers alike.

EVALUATION:

Presentation—Fair to Good
Rules—Good
Playability—Very Good
Realism—Fair to Good
Complexity—3

OVERALL EVALUATION: Good

La Bataille d'Auerstädt (1978)

PUBLISHER: Marshal Games

SUGGESTED RETAIL PRICE: $13 (resealable plastic bag)

SUBJECT: The Battle of Auerstädt in 1806.

PLAYING TIME: The game takes twelve to fifteen hours to complete.

SCALE: This is a grand-tactical game. Most units are regiments. Each turn equals twenty minutes; each hex represents 125 yards.

SIZE: The 22″ x 34″ game map uses 16mm (⅝″) hexes.

BALANCE: Balance is not bad, although the sympathies of the designers are obviously with Napoleon, and the better player should take the Prussians.

KEY FEATURES: The game is a direct descendant of the exceptional game *La Bataille de la Moskowa* and uses most of its rules intact (if somewhat mangled). The sequence is fairly complex: Charge/Movement/Defensive Fire/Offensive Fire/Melee/Morale Check. All units have several ratings, usually with large double-digit numbers. The emphasis here is on firepower and morale. Formations are used, and leaders can be a factor.

COMMENTS: This is a strange game. The counters are beautifully printed; yet players will find it difficult to read them. The rules are written in a cutesy, fractured Franglais that brings Inspector Clouseau to mind. The rules should make sense, but they don't. When they do, they can be remarkably dumb. Even so, the feel for Napoleonics is rather good. Many of this game's excesses have been corrected in later games using the same system. Still, a magnificent production job is marred by a poorly developed and badly explained system. It's for diehards.

EVALUATION:

Presentation—Very Good
Rules—Ridiculous
Playability—Poor
Realism—Very Good
Complexity—8

OVERALL EVALUATION: Poor

La Grande Armée (1972)

PUBLISHER: Simulations Publications, Inc.

SUGGESTED RETAIL PRICE: $12 (plastic box)

SUBJECT: The land campaigns of Napoleon in Central Europe, 1805–1809.

PLAYING TIME: Each of the game's three scenarios can be played in an evening.

SCALE: This is an operational game using a variety of formation levels from army to brigade. Each hex represents 9.4 miles; each turn represents ten days.

SIZE: The 22″ x 34″ uses 16mm (5/8″) hexes.

BALANCE: There is perhaps a slight edge given to the French.

KEY FEATURES: This is quite an unusual system that is rarely copied. Movement allows forced and triple-forced marches, with concomitant disruptions and stragglers. Movement also depends on whether you want to move units as armies, corps, divisions, or whatever; the smaller the unit, the faster it moves, but it costs you to break it down and build it up. Zones of control depend on the type of unit and terrain, but they are rigid. Combat is voluntary, and the odds/ratio Combat Results Table results in eliminations, exchanges, or "scattered" retreats (in which the units must break down into the smallest format). Supply—using moving wagon trains—is very important. Leader rules are designed so that good leaders such as Napoleon can make or break a battle. Victory is by geographic objectives.

COMMENTS: If there were such a thing as an SPI classic, this would be one. It's nearly ten years old but is still one of the most playable and—in terms of design intent—most accurate of Napoleonic games. It's fairly easy to play. There are few charts and relatively straightforward ideas. The system places a premium on the Napoleonic ideal of dispersal for movement and concentration for battle. The game moves very fast, and situations develop with remarkable speed. *La Grande Armée* requires good bluffing powers and steady nerves. Despite its stark appearance, it's a top item.

EVALUATION:

Presentation—Good
Rules—Good
Playability—Very Good
Realism—Good to Very Good
Complexity—6

OVERALL EVALUATION: Very Good

Napoleon (1977—second edition of a game originally published by Gamma Two Games, Ltd., in 1974)

PUBLISHER: Avalon Hill Game Co.

SUGGESTED RETAIL PRICE: $12 (boxed)

SUBJECT: Napoleon's campaign in Belgium against the Anglo-Prussian Allies in June, 1815.

PLAYING TIME: The game takes one to two hours.

SCALE: This is a strategic or operational game (take your pick). The basic unit is the corps, but time and distance scales are fairly abstract. Hexes are not used on the game boards.

SIZE: Two 11″ x 16″ boards form a 16″ x 22″ playing surface. A separate 8″ x 32″ area is needed to set up two battle cards.

BALANCE: Good.

KEY FEATURES: *Napoleon* employs a variation of the unusual system introduced in *Quebec, 1759* (reviewed in Chapter 8). The domino-like units move from town to town along a network of roads instead of from hex to hex. Movement limitations depend on whether the road traveled is a major or minor road and whether any bridges must be crossed. Units must be concentrated into groups, and each player may move only two groups each turn. No Combat Results Table is used; instead, units roll one die per strength point. A rolled 6 causes the loss of an enemy strength point. Units face off in three columns and fire on each other until one player is forced to vacate a column. Movement of reserves is detailed as is reinforcement from nearby towns.

COMMENTS: *Napoleon* is a fresh experience for the land-combat enthusiast. Although it doesn't resemble a simulation very much, it is a challenging game that can be fun to play. The difficulties presented by the movement system are far removed from those in more typical wargames. The vagaries of the dice produce scrambling, tension-filled battles in which a decisive defeat can be turned into victory in an amazingly short time. *Napoleon,* however, is not merely a die-rolling contest. The real emphasis is on concentrating sufficient force to overcome any freak occurrence. It's a refreshing break from the rigors and eyestrain of more conventional treatments.

EVALUATION:
Presentation—Very Good
Rules—Good
Playability—Excellent
Realism—Fair
Complexity—4

OVERALL EVALUATION: Good to Very Good

Napoleon at Leipzig (1979)

PUBLISHER: Operational Studies Group

SUGGESTED RETAIL PRICE: $18 (boxed)

SUBJECT: The Battle of Nations (one of the largest land battles ever), October 16–19, 1813.

PLAYING TIME: The three scenarios take an evening or longer to complete. An entire campaign game will take a weekend.

SCALE: This is a battle-level game with operational flavor. Units are mostly brigades. Each turn represents an hour except at night, and each hex equals 480 meters.

SIZE: The two 22″ x 34″ game maps use 16mm (5/8″) hexes and require a three-by-four-foot playing surface.

BALANCE: Balance seems to favor the French.

KEY FEATURES: The system is adapted directly from Simulations Publications' *Napoleon's Last Battles* (unsurprisingly, since the designer of the quad—Kevin Zucker—is now one of the principals at Operational Studies Group). However, much has been added to the system, especially in the way of command control. The Campaign Game and Grand Tactical Game introduce a whole series of command rules and personality profiles: leaders must be able to trace "orders" to their units; combat units need leaders to enter an enemy zone of control; leaders reorganize shattered units, and so on. Additional rules cover demoralization (the basis of victory), types of orders, divisional integration, canister, initiative, reserves, cavalry charges, and Congreve rockets.

COMMENTS: This is a stunningly beautiful game. From the box cover to the spectacular map and the gaily colored counters (which are, perhaps, a bit too busy), everything is calculated to please the eye. The game also includes excellent historical notes and a good order of battle. The game is based on a time-honored system and, as such, plays well. The added command features are interesting, but they do make things top-heavy in that department for what is otherwise a fairly simple system. The main problem is in balance, as the French seem to do a bit too well. Whether this is caused by the system or design bias is difficult to determine, but it takes the edge off what is otherwise a very nice effort.

EVALUATION:
Presentation—Excellent
Rules—Good (some pages are reversed)
Playability—Good
Realism—Good
Complexity—Basic Game: 5; Advanced Rules: 7

OVERALL EVALUATION: Good

Napoleon at War QuadriGame (1975—includes the folio games *Marengo*, *Jena-Auerstädt*, *Wagram*, and *The Battle of Nations*)

PUBLISHER: Simulations Publications, Inc.

SUGGESTED RETAIL PRICE: $14 (plastic box; individual folio game, $4)

SUBJECT: Four land battles from the 1800–1813 period.

PLAYING TIME: One to four hours, depending on the battle.

SCALE: These are operational simulations. Depending on the battle, units are demi-brigades to divisions; hexes are four hundred to eight hundred meters, and turns represent one to two hours.

SIZE: The four 16″ x 20″ maps (one per battle) use 16mm (5/8″) hexes.

BALANCE: This is reasonably close in all games.

KEY FEATURES: All four games in the quad use the basic *Napoleon at Waterloo* system. In addition to a standard, shared set of rules, each game has a short set of rules that are unique to the game. Cavalry is distinguished from infantry only in being faster and, usually, in smaller units. Since the Combat Results Table is relatively bloodless, and the zones of control "locking," engaged units tend to stay engaged for a long time. All games except *Marengo* have a demoralization rule. This makes it difficult to conduct attacks when demoralized. Terrain effects vary with each game (although they are usually important), as do victory conditions.

COMMENTS: The *Napoleon at War* Quadri-Game is a fine set of game player's games (even SPI does this sort of thing every once in a while). Mechanics are simple and easily learned, and the games are clean and fast moving. The four can almost be considered as scenarios of one large game. Their simplicity and short length makes them good introductory games. Their greatest weakness, perhaps, is that they don't have many strategic options, so they may get stale after you've played them a few times. Among the four, however, they offer a good bit of game playing. *Marengo* is the smallest, shortest —and best. It provides both players with good offensive opportunities, and the battle is generally not decided until the final attack. *Jena-Auerstädt* is probably the least successful. The Prussian player has the option to fight the battle in the historical way or to concentrate his entire army at Jena against Napoleon. In the first case, a fairly dull battle ensues on both map segments. In the second, only the Jena battle is fought, which is more interesting, but it's still slower moving than the other games in the quad. *The Battle of Nations* is larger than the others and offers three scenarios based on playing all or part of the battle.

EVALUATION:
Presentation—Good
Rules—Good
Playability—Excellent
Realism—Fair to Good
Complexity—3

OVERALL EVALUATION: Very Good

Napoleon at Waterloo (1971)

PUBLISHER: Simulations Publications, Inc.

SUGGESTED RETAIL PRICE: $12 (plastic box)

SUBJECT: The Battle of Waterloo in June, 1815.

PLAYING TIME: One to two hours are needed for the Basic Game; two to three hours for the Advanced Game.

SCALE: This is an operational-level simulation. In the Basic Game, units are divisions; in the Advanced Game, units are brigades.

SIZE: The 17″ x 23″ game map uses 16mm (5/8″) hexes.

BALANCE: The Allies have a slight edge in the Basic Game, but this is reversed in the Advanced Game.

KEY FEATURES: *Napoleon at Waterloo* features a straightforward game system intended for introducing novices into the hobby; it's the basis for many of SPI's other Napoleonic (and some other) games. The Basic Game is very simple standard stuff. Artillery has ranged-fire capabilities, and morale for both sides is a function of cumulative losses. The French win by exiting units toward Brussels, and the Allies win by preventing this. The Advanced Game uses a different counter mix and adds various rules and modifications. Artillery fire separately, before infantry attacks, and cavalry can move through zones of control (with increased movement costs). Combat is optional and includes various die-roll modifications, depending on the units involved. Terrain and morale are more involved.

COMMENTS: The basic version of *Napoleon at Waterloo* was formerly sent to *Strategy & Tactics* magazine subscribers as an introduction to wargaming. It is quite good for that, and arguably better in its way than the Advanced Game. Despite its simplicity, a good deal of tension is generated during play. The advanced version's complications are simple enough so that novices should be able to assimilate them in minutes after having played the Basic Game. The total package is a fine two-step introduction to Napoleonic wargames.

EVALUATION:
Presentation—Fair to Good
Rules—Very Good
Playability—Excellent (Basic Game) to Very Good (Advanced Game)
Realism—Fair to Good

Complexity—3
OVERALL EVALUATION: Good to Very Good

Napoleon's Last Battles QuadriGame (1976—includes the folio games *Ligny, Wavre, Quatre Bras,* and *La Belle Alliance*)

PUBLISHER: Simulations Publications, Inc.

SUGGESTED RETAIL PRICE: $14 (plastic box; individual folio game, $4)

SUBJECT: The Waterloo campaign in June, 1815.

PLAYING TIME: Two to three hours per folio game; six to ten hours for the campaign game.

SCALE: This is an operational-level simulation. Most units are brigades. Hexes are 480 meters across, and a turn equals one hour.

SIZE: The four 22″ x 17″ maps (one for each folio game) use 16mm (⅝″) hexes. When used together for the campaign game, they form a 43″ x 33″ playing surface, which may tax the average game table.

BALANCE: This varies a bit but is generally good.

KEY FEATURES: This group of games uses the tried-and-true *Napoleon at Waterloo* system with some elaboration. Artillery may bombard from a distance, and line of sight is taken into account. There are rules for combined-arms attacks, the Imperial Guard, as well as special rules for the individual battles. Morale rules are included, and "disintegration"—one step down from demoralization—is introduced. While objectives vary from game to game, the key to victory in all games is demoralizing the opposing army. The four games (and their maps) combine into one large battle game with special rules that introduce the elements of command, reorganization, and supply. The main additions to the old system are the supply rules and the command structure, which severely limit movement of units.

COMMENTS: *Napoleon's Last Battles* is one of SPI's most popular and highly rated games. (The system has been adapted wholeheartedly for Operational Studies Group's Napoleonic series: see *Napoleon at Leipzig.*) While similar to the *Napoleon at War* quad, these games are less enjoyable. This isn't due to the changes in the game system, which are minor, but to the scope of the battles involved. Whereas the *Napoleon at War* battles are open and free-ranging, the battles in *Napoleon's Last Battles*—especially *Ligny* and *La Belle Alliance*—are set-piece affairs in which the object is to wear the enemy down to his demoralization level, a not particularly fascinating prospect. The campaign game is aimed at a somewhat different audience—one happier with longer and more complicated games—but the scale is wrong for a game of the Waterloo campaign. The standard rules in two colors are hard to look at, let alone read.

EVALUATION:
Presentation—Very Good
Rules—Good
Playability—Excellent (except for campaign game)
Realism—Good
Complexity—Folio Games: 4; Campaign Game: 6
OVERALL EVALUATION: Good to Very Good

Ney vs. Wellington (1979)

PUBLISHER: Simulations Publications, Inc.

SUGGESTED RETAIL PRICE: $10 (boxed)

SUBJECT: The Battle of Quatre Bras, June 16, 1815.

PLAYING TIME: The game takes four to eight hours.

SCALE: This is a tactical-level simulation using battalions and regiments. Each hex represents one hundred yards, and each turn equals fifteen minutes.

SIZE: The 22" x 34" game map uses 16mm (5/8") hexes.

BALANCE: This is reasonably close.

KEY FEATURES: The game uses an unusual sequence of play: a mutual fire phase occurs between the two player turns, while shock attacks take place during the opposing player's turn. Formation rules have appropriate effects on movement, stacking, and combat. Other variables in the two types of combat include firing strength and effectiveness ratings, which are far more important than unit strengths and are reduced as casualties are taken. Leaders are important for morale considerations. There are special rules for cavalry and artillery.

COMMENTS: The system here is adapted from *Wellington's Victory*, and the sequence of play may well be initially confusing. The vast importance of effectiveness ratings—as opposed to actual combat strengths—can also be a source of dismay for gamers who are less accustomed to the peculiarities of some modern Napoleonic simulations. The line-of-sight algorithm used at times requires division to two or more decimal places. A line-of-sight chart is supplied, but an inordinate amount of time is required to resolve each situation. The recent trend at SPI toward step reduction requires two counters for virtually every unit, which can be annoying. There is a major shortage of strength counters provided in the counter mix, and as they are vital to game functions, you must be prepared to manufacture suitable counters on your own or scavenge them from other games. Despite these shortcomings, *Ney vs. Wellington* is a fine simulation: detailed, challenging, and educational.

EVALUATION:

Presentation—Good to Very Good
Rules—Good
Playability—Fair
Realism—Very Good
Complexity—8

OVERALL EVALUATION: Fair to Good

Waterloo (1962)

PUBLISHER: Avalon Hill Game Co.

SUGGESTED RETAIL PRICE: $12 (boxed)

SUBJECT: The Battle of Waterloo in June, 1815.

PLAYING TIME: The game takes about four hours.

SCALE: This is an operational-level simulation. The basic unit is the division. Hexes equal five hundred yards, and turns represent two hours.

SIZE: The 22" x 28" game board has 16mm (5/8") hexes.

BALANCE: This is good.

KEY FEATURES: *Waterloo* uses the original Avalon Hill formula (move/fight, odds/ratio Combat Results Table, and so on) as applied to the Napoleonic era. There is no differentiation of infantry, cavalry, and artillery other than speed of movement. Players may stack up to fifteen combat factors in a hex, regardless of the number of units involved. Victory for the French is attained by eliminating the Prussian and

Waterloo, a game of the Napoleonic era from Avalon Hill. (PHOTO: AVALON HILL GAME CO.)

Anglo-Allied armies through combat or defection—the latter controlled by French geographical objectives.

COMMENTS: *Waterloo* is the spiritual progenitor of all of the operational Napoleonic-era simulations. It is first and foremost a *game,* and French combat factors were modified for play balance. Much of the terrain is imaginary. While it is seriously lacking in realism, the game captures something of the feel of the period and remains fun to play.

EVALUATION:
- Presentation—Good
- Rules—Very Good
- Playability—Very Good
- Realism—Poor
- Complexity—3

OVERALL EVALUATION: Good

CHAPTER 10

The Dawn of Modern Warfare

For the military historian, the mid-nineteenth century is one of the great seminal periods, the beginning of what we know as modern warfare. The Napoleonic era had formalized the tactics that had evolved during the years prior to the Emperor's ascendancy, and by the time of Waterloo the great generals and military thinkers had perfected a style of warfare that, broadly speaking, had endured for almost two centuries. When, after several decades of relative peace, war again reared its ugly head, these same generals—and this same trend of thought—were still paramount.

The main aim of Napoleonic warfare, the logical successor to the tactics of Marlborough, Frederick, and other eighteenth-century generals, was to close with the enemy and drive him away. To this end artillery, which wasn't enormously advanced over the cannon of the seventeeth century, was used to soften up troops for a mass assault. The infantryman's musket delivered a single ball rather erratically over a limited range (not much in excess of one hundred yards) and took some time to reload. The final weapon was still shock: the clash of men hurling themselves against other men in much the same way that Caesar's legions attacked their foes two millennia previously.

Ruling over the final assault was the cavalry charge. Given its ferocity and power, against which infantry might get off no more than a round or two before the horses were on them, it's not surprising that many formations of troops were designed specifically against the mounted charge. Thus, Napoleonic tactics depended on fields of skirmishers to impede masses of infantry and densely packed but unwieldy defensive formations like the square to counter cavalry.

By the time of the American Civil War, virtually all of this was outmoded. Technology had far outstripped military philosophy, and the result was a series of horrible lessons slowly learned by stubborn and often inept generals—lessons not fully brought home until the horror of World War I. If we can dismiss Custer and the Indian Wars as a form of guerilla warfare somewhat ahead of its time, we can discern from the games based on this period a steady progression of military technology centering mainly on U.S. affairs, running from Alma to Buena Vista, through Gettysburg, and ending with Sedan and Metz.

Only two conflagrations in this era are covered with any completeness in wargames: the Crimean War and the American Civil War. Games on relatively minor conflicts like the Mexican War and the Indian

Wars (limited, in wargaming, to Custer's debacle at the Little Big Horn) are rare, and simulations of other European, African, Asian, or South American conflicts are rarer still. This is the only historical era so dominated by U.S. history. While the Civil War was one of the most cataclysmic events of the period—militarily, socially, and politically—its existence doesn't entirely excuse the neglect of a more global outlook by game designers.

Were it not so horrible an operation, the Crimean War would stand as a Woody Allen version of warfare. In retrospect, it's difficult to believe the stupidity and obstinacy of those involved. Supply was primitive, not because the generals failed to appreciate railroads, but because they ignored them. They fought the war in the Crimea, an area almost inaccessible to steady supply. In fact, most of the conflict consisted of a protracted and poorly handled siege of Sevastopol, interrupted by cholera and occasional Russian attempts to relieve the city and reopen supply lines. The conduct of the rest of the war still relied heavily on Napoleonic tactics, which were not entirely outmoded because good rifled weapons were not yet readily available.

The Crimean War itself is covered quite well in simulation, although it wasn't until a few years ago that the state of the art—and the sophistication of the gamers—could handle the mechanics necessary to reproduce such a strange conflict. In its concentration on supply and command, along with the rules needed to cover the subject properly, *Crimea* from Game Designers' Workshop has admirably captured the feel of the campaign. Although it is now several years old, *Crimea* is a fairly advanced game in that it attempts to treat the topic in an intelligently operational manner, overlooking little, rather than reducing the campaign to one viewpoint, as earlier games often did. Concentrating on the battles of the war, Simulations Publications' *Crimean War* QuadriGame uses another fairly sophisticated system. While it emphasizes the reliance on Napoleonic tactics, it manages to show how those tactics were becoming outmoded.

It's a well accepted conclusion that the American Civil War was the first "modern" war in the sense that it was the first time men were moved rapidly from one theater to another by means of railroads, a fact that changed the operational face of warfare forever. No longer were large masses of infantry limited by their speed afoot; nor did an army have to march—slowly—with its entire baggage and supplies as long as it had a network of rails behind it.

Even more modern was the fact that in the Civil War men began killing from a distance. Rapid advances in the use of steel had made cannon more accurate and lethal, but it was the rifled musket that brought about the revolution. The killing range of firearms was extended to about three times their previous limit, from about one hundred fifty yards to over four hundred yards. Men had three times as long—and about three times as many shots—to halt a charging foe. Moreover, rifling the barrels of muskets made greater accuracy possible. By 1865, at the end of the war, breech-loading rifles had replaced muzzle-loading muskets so that the average soldier had ten times the killing power of a member of Napoleon's *La Grande Armée*.

Because of this, cavalry charges became doomed affairs that were rarely attempted, and "human wave" assaults proved to be bloody and fruitless efforts. It was the Civil War—in the Hornet's Nest at Shiloh, at Mayre's Heights in Fredericksburg, and at The Angle (against Pickett's Charge) at Gettysburg—that brought home the lesson that a hundred men defending a well entrenched position could easily hold off three times their number.

Few events are simulated better—or in greater variety—than the American Civil War. It's hard to name a company that doesn't have some sort of a game on the subject. In fact, Avalon Hill built much of its early product line on the Civil War, with the now-defunct (and very strange) *Civil War, Chancellorsville* (revised in 1976), and *Gettysburg* (which appeared in a host of different editions, finally culminating in the all-new "triple" game commonly known as *Gettysburg '77*). None of these games, however, treated the subject on anything higher than a "fun" level, because, regardless of the version, *Gettysburg*—as played—rarely looked much like the actual battle.

The first attempt at a serious simulation of the Civil War was SPI's magazine game, *American Civil War.* The entire war was replayed on a highly strategic level, using a unique—but rather confusing and tiring—system. Although not unpopular, the game had little flavor (and several fluffs with the railway lines). Undaunted, SPI followed with *Wilderness Campaign* and *Lee Moves North* (originally entitled *Lee at Gettysburg*). Both were improvements of a simultaneous movement system pioneered in SPI's discontinued *Franco-Prussian War. Lee at Gettysburg* was the ultimate manifestation of this system and, despite its age, remains one of the best simulations of Civil War operations ever devised. Although it lacks enough actual combat for the tastes of most gamers, it captures the flavor of campaigning succinctly.

A lull in the gaming world over the next several years was broken only by a pair of somewhat misguided, and no longer available, attempts to render the Civil War on a tactical level: SPI's *Rifle & Saber* and Simulations Design Corporation's *Rifle & Musket.* Battles were pretty much left to the beer-and-pretzels format, a policy seemingly reinforced by the success of the *Blue & Gray* QuadriGames. The feeling seemed to be that the Civil War, as a serious matter, was only for quaint historians.

Then came *Terrible Swift Sword,* one of the most popular games SPI has ever produced. With one stroke—and a giant one at that—Civil War games reached maturity. This incisive simulation opened the floodgates, and such serious efforts as *Shenandoah, Objective: Atlanta,* and *War Between the States* came pouring out. While intended only for experienced—and patient—players, such games finally allowed the enthusiast and historian of the period to play not just for fun, but to learn something.

For Civil War buffs the future looks exceedingly bright. SPI was preparing to bring out both giant (like *Bloody April*) and smaller games (like *Stonewall*) that are based on the *Terrible Swift Sword* system, and several new companies are bringing out their own offerings.

There are, however, many major military events of this period that are not covered by gaming. The Taiping rebellion against the Manchu dynasty of China, one of the most vicious and bloody wars of all time, surely cries out for simulation. The Sepoy rebellion in India—a war in which the word "prisoner" was unknown—has also been ignored. South America, in constant turmoil during these years, might as well be invisible for all the gaming world sees of it.

Some of these oversights are sure to be corrected because this is a fascinating era, one that saw the burst of the Industrial Age together with an explosion on all levels of technology. It hailed the dawn of the modern age, and it's an era of rare opportunity for game publishers.

Evaluations

Blue & Gray QuadriGame (1975—includes the folio games *Shiloh, Antietam, Cemetery Hill,* and *Chickamauga*)

PUBLISHER: Simulations Publications, Inc.

SUGGESTED RETAIL PRICE: $14 (in plastic box; $4 per individual folio game)

SUBJECT: Four land battles in the American Civil War, 1862–1863.

PLAYING TIME: The game takes one to four hours to play.

SCALE: These are grand-tactical games. Units are mostly brigades, but some Union infantry units are divisions. Each hex represents a distance of four hundred meters, and each turn represents one to four hours, depending on the individual game.

SIZE: The four 22″ x 17″ maps (one for each game) use 19mm (¾″) hexes.

BALANCE: This varies. The Confederate forces are favored in *Chickamauga* and the Union, to a lesser degree, in *Shiloh*. The other two games are fairly even.

KEY FEATURES: *Blue & Gray* utilizes the *Napoleon at Waterloo* game system, the mechanics of which will be familiar to any gamer with even modest experience. There is no functional difference between infantry and cavalry, but artillery has ranged fire capabilities. The odds/ratio Combat Results Table is heavy on retreats. Special rules govern night turns and restrict movement of engaged units. Options include unit-effectiveness rules simulating the wear and tear on attacking units. There are also a few special rules for each folio game.

COMMENTS: The simple rules and mechanics of *Blue & Gray* are ideal for short playing times and for introducing newcomers to wargaming. The games are also excellent for solitaire play. Unfortunately, only *Chickamauga* is really fluid; the other three games are rather lacking in movement. Movement restrictions, which are used to simulate Union indecisiveness, hamper the Union player in *Shiloh* and *Antietam*. While perhaps necessary for balance, they're tiresome. Use of the optional combat-effectiveness rules swings the changes of victory heavily in favor of the defender and certainly doesn't help the excitement level. These are solid but unspectacular games suitable for a casual afternoon of beer and pretzels. The graphics are colorfully done.

EVALUATION:

Presentation—Very Good
Rules—Good
Playability—Very Good

Realism—Good
Complexity—3
OVERALL EVALUATION: Good

Blue & Gray II QuadriGame (1975—includes the folio games *Fredericksburg*, *Chattanooga*, *Hooker and Lee*, and *Battle of the Wilderness*)
PUBLISHER: Simulations Publications, Inc.
SUGGESTED RETAIL PRICE: $14 (in plastic box; $4 per individual folio game)
SUBJECT: Four land battles in the American Civil War, 1862–1864.
COMMENTS: The evaluations and comments on the previous *Blue & Gray* QuadriGame apply here, too. The battles are fairly comparable, although none quite matches *Chickamauga* for interest. *Battle of the Wilderness* features the only fluid situation, but even there the Union player must contend with movement restrictions imposed by the rules. *Hooker and Lee* allows an off-map flanking maneuver by the Confederate forces; otherwise, it is static. The other two games are assaults on prepared defensive positions. The Grand Chancellorsville Option, which enables players to combine the *Fredericksburg* and *Hooker and Lee* games into one larger campaign, could serve nicely as the second step in a novice's introduction to the hobby. As a whole, this is a handy but not essential group, and it's not as attractive as *Napoleon at War*.

Chancellorsville (1961, 1974)
PUBLISHER: Avalon Hill Game Co.
SUGGESTED RETAIL PRICE: $12 (boxed)
SUBJECT: The American Civil War battle between Hooker and Lee in May, 1863.
PLAYING TIME: An evening is needed to play.
SCALE: This grand-tactical battle game is more operational in feel than anything else. Most units are divisions (Union) or brigades (Confederate States of America). Each hex covers one mile, and each turn represents three hours.
SIZE: The 22″ x 28″ game map uses 16mm (¾″) hexes.
BALANCE: This isn't good; the Union player should win at least three out of five games.
KEY FEATURES: the sequence of play is standard Move/Fight, with alternating artillery fire phases thrown in. Except for the pontoon bridges, movement is also simple. Zones of control are rigid, and combat is mandatory. The regular Combat Results Table is odds/ratio and yields mostly retreats and disruptions. There are some additional rules (albeit simple ones) dealing with command control, Stoneman's Raid, inverted Confederate units to aid play balance, and so on. This new version is not too different from the old one, although much has been polished and refined.
COMMENTS: *Chancellorsville* is an Avalon Hill pseudoclassic. Do you remember the old Avalon Hill "classics"—games that were lots of fun to play but bore as much resemblance to history as the Toronto Blue Jays do to a baseball team? *Chancellorsville* qualifies on the latter count but fails on the former; it isn't much fun to play, either. Chancellorsville was a large, complex battle. Reducing it to a scale of one mile to the hex was a mistake because it makes no sense at this level. Furthermore, the tremendous problems inherent in the battle are largely ignored by the de-

signers: Hooker's indecision and inactivity and Jackson's flank maneuver are all but impossible. The differences between the two armies—the differences that decided the battle—are swept under the rug of overt simplicity. The map is also not the best. It's better than it used to be back in 1961, but that's not saying much.

EVALUATION:
Presentation—Fair to Good
Rules—Very Good
Playability—Very Good
Realism—Poor
Complexity—4

OVERALL EVALUATION: Fair

Crimea (1975)

PUBLISHER: Game Designers' Workshop

SUGGESTED RETAIL PRICE: $10 (resealable plastic bag)

SUBJECT: The game covers the first year of the Crimean War (1854–1856) in Russia between Britain, France, and assorted allies and their common opponent, Russia. Minor naval aspects are included.

PLAYING TIME: One hour is needed for the (very) short scenarios; about twenty-four hours for the complete campaign game.

SCALE: This is an operational game using brigades and regiments. Each turn covers two weeks.

SIZE: The 22″ x 28″ game map uses 16mm (⅝″) hexes.

BALANCE: The better player should take the Russian side, but, since winning depends less on combat and more on endurance and cunning (in the campaign, at least), balance depends more on player personality than on the rules.

KEY FEATURES: This is a unique game; nothing like it has been done before—or since. The map covers the four different areas where events took place; these are linked by a communications map. Although the rules are deceptively short, there are lots of charts and order-of-battle sheets. The sequence of play is Fortification Building/Strategic Movement/Action Segment (combat, if any). The Action Segment, which reproduces the battles on a grand-tactical level (an unusual approach), is a game within the game. Units hold fire, move out of harm's way, use siege fire, repair fortifications, fire again, then melee. Morale is checked often during battle, and the use of reserves is quite important. Movement on roads is virtually unlimited but quite restricted off them. Much of the game involves setting up and protecting lines of communication. Optional rules add the use of navies et al.

COMMENTS: This is a game you just want to like (unless, of course, you're among those who think the world began when Hitler attacked Poland). Except for the drab map, the game is very attractive, and the system is intriguing. It also depicts the campaign fairly accurately. As with many games in which a system has been designed to form-fit the subject, the problem is that the game isn't really a game. There's too much jockeying for position (as opposed to maneuvering for position; the difference is that the former is more "faint" than feint). And, like history, the game bogs down around Sevastopol, which the Allies reach quickly, unless they are totally incompetent. Then it becomes an operational siege game, with some random attacks. It is an interesting design, though, and for that it may be worth inspection.

EVALUATION:
Presentation—Very Good
Rules—Fair
Playability—Good
Realism—Very Good
Complexity—8
OVERALL EVALUATION: Fair to Good

Custer's Land Stand (1976)
PUBLISHER: Battleline Publications
SUGGESTED RETAIL PRICE: $10 (boxed)
SUBJECT: Nominally, it's the Battle of the Little Big Horn, June 25, 1876; actually, it's the campaign of the 7th Cavalry against the Sioux and other Indians.
PLAYING TIME: The game takes about five hours.
SCALE: This is an operational game using cavalry companies and Indian war parties (divided into subtribes). The scale appears to be about four hundred yards per hex; each turn represents about an hour.
SIZE: The 22″ x 27″ game map uses 16mm (5/8″) hexes.
BALANCE: The Indians, naturally, will win more than half the time, but Custer has a fair chance.
KEY FEATURES: This is a fairly simple system with some wrinkles. Movement and combat are straightforward; the odds/ratio Combat Results Table yields eliminations, retreats, and exchanges. Gatling guns, cavalry charges, and pack trains provide combat adjustments. There are some interesting rules for scouting and spotting the Indians or cavalry. A host of optional rules—Benteen's troops, artillery, and so on—provides further "what-if" grist for the players' mill.
COMMENTS: There may have been more lopsided events in military history, but not many spring to mind. So why a game? Simple: The year 1976 was the Custer centennial! Hence, there were several efforts on the subject, among them this one. The game is not badly done, but the situation is worth about one play; after that it becomes a bit stale. Most of the fun is in avoiding being spotted by the enemy; after that it simply becomes a free-for-all. The treatment, however, is probably as good as could be expected, but the situation itself is so bad that it drags the game down. A curious but stolid affair.
EVALUATION:
Presentation—Good
Rules—Very Good
Playability—Good
Realism—Good
Complexity—5
OVERALL EVALUATION: Fair

Fury in the West (1977)
PUBLISHER: Battleline Publications
SUGGESTED RETAIL PRICE: $10 (boxed)
SUBJECT: The Civil War Battle of Shiloh in April, 1862.
PLAYING TIME: A long evening is required.
SCALE: This is a battle-level game that is not truly tactical in feeling. The units are brigades; each turn represents an hour; and each hex appears to represent about 450 yards.
SIZE: The 22″ x 27″ game map uses large 25mm (1″) hexes.
BALANCE: The Union has an edge here, but a wily, determined Confederate player can prevail.
KEY FEATURES: The infantry brigades are double-sized counters; however, as each hex is quite large, they are still treated pragmatically as regular-sized counters.

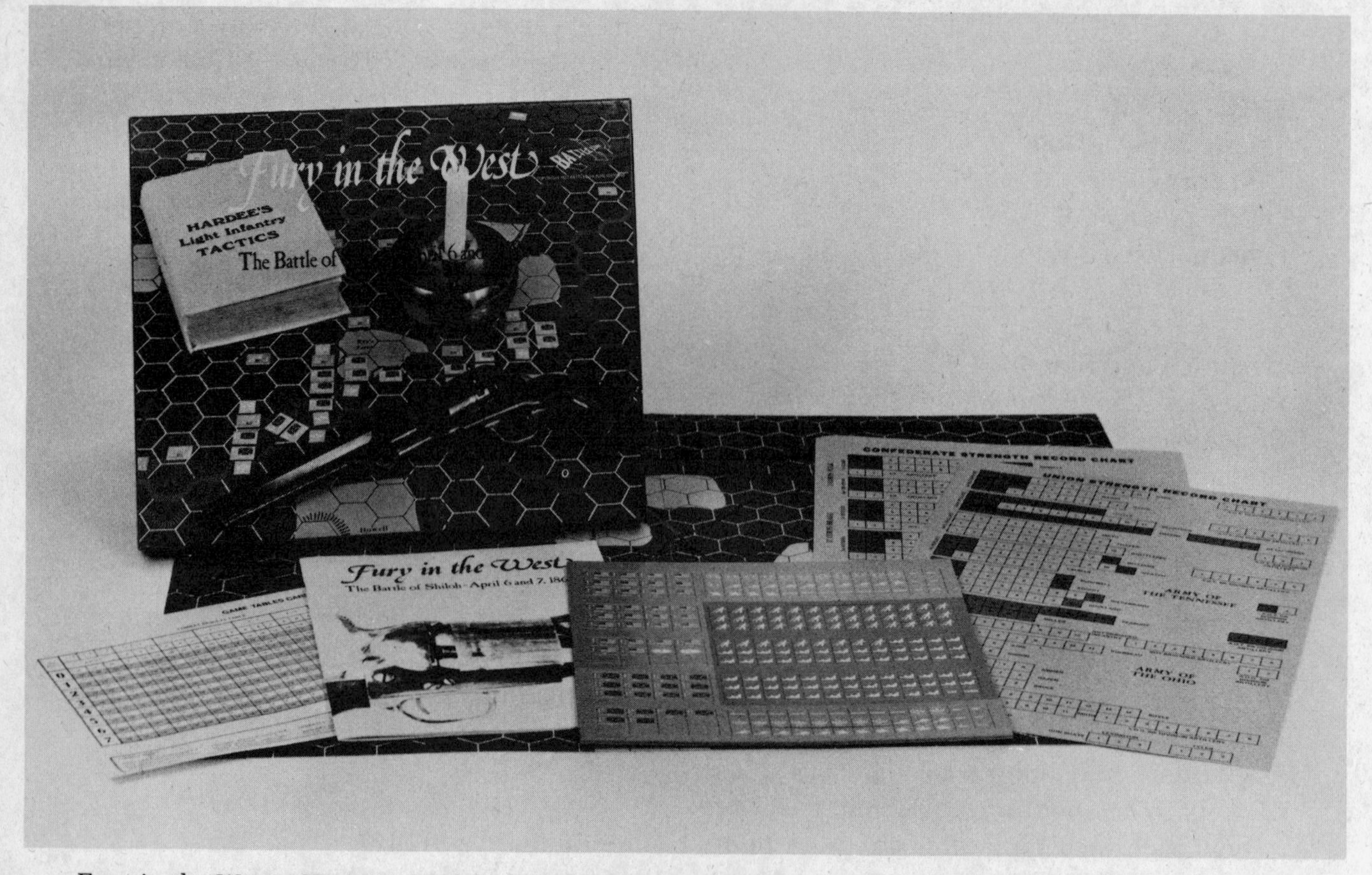

Fury in the West—*Heritage Models' board game of the Battle of Shiloh, 1862. A realistic historical strategy game that depicts a crucial Civil War battle. This is one of Heritage's many "Battleline" Games for the adult (12 years and up).* (PHOTO: HERITAGE MODELS, INC.)

There is an emphasis on facing and stragglers; units lose stragglers each time they move or retreat. The sequence of play is quite simple: Move/Fight, then Move/Fight. The Combat Results Table is odds/ratio but with different columns depending on terrain—which, being mostly wooded, is very important. Results are in losses, routs, withdrawals, and so forth. Artillery and cavalry are covered by advanced or optional rules, as are formations, bayonet charges, and gunboats.

COMMENTS: Designer Steve Peek has done some good games on the Civil War (like *Shenandoah*), but let's face it: This is a buzzard—at least as a simulation. As a game, it's not bad, because Shiloh can be a nip-and-tuck affair until the end of the first day—which is what this game covers. But forget historicity. The map is a travesty (the key Sunken Road, the backbone of the formation of the Hornet's Nest, should be *north* of the Peach Orchard, not south of it!), and the key rule—stragglers—uses the wrong means to achieve its ends. Yes, there was a great deal of straggling at Shiloh, but little, if any, resulted simply from maneuver. Most of it occurred from combat or looting, and it's ignored here. In *Fury in the West*, every time a unit moves it loses stragglers: sheer folly! Some of the unit strengths are questionable, and morale, a key ingredient of the battle, is missing. It plays well, though.

EVALUATION:
- Presentation—Very Good
- Rules—Good
- Playability—Good
- Realism—Fair to Poor
- Complexity—5

OVERALL EVALUATION: Fair

Gettysburg (1977—sometimes known as *Gettysburg '77* to distinguish it from earlier versions)

PUBLISHER: Avalon Hill Game Co.

SUGGESTED RETAIL PRICE: $12 (boxed)

SUBJECT: The Civil War Battle of Gettysburg in 1863.

PLAYING TIME: It takes two to four hours for the Basic Game, eight hours for the Intermediate Game, and forty-eight hours for the Advanced Game.

SCALE: This is a grand-tactical game using divisions (Basic Game) down to regiments (Advanced Game). Each hex equals 756 feet. The turn equivalent varies: it's four hours for the Basic, one hour for the Intermediate, and twenty minutes for the Advanced Game.

SIZE: The 22″ x 28″ game map uses 16mm (5/8″) hexes.

BALANCE: Thanks to hindsight that Lee didn't have, balance is slightly in favor of the South.

KEY FEATURES: There are three different games on the same battle here, and each has nothing to do with the others. The Basic Game is wargaming at its simplest: pure Move/Fight, DE/AE stuff. Terrain is ignored. Most of the Intermediate Game is also familiar; it has a slightly more complex Combat Results Table and rules covering breastworks, disorder, multiple rounds of combat, and command structure. The Advanced Game is something else. It uses a host of complex rules—extended battlelines, formations, facing, line of sight, ammunition—as well as a sophisticated activity-and-movement system that allows the players a remarkably wide range of choices. Units can overextend themselves for a major assault with the probability that they will become disorganized and fatigued quite rapidly. Again, although line-of-sight considerations are common, terrain is virtually ignored for movement.

COMMENTS: The Basic Game is no more than a learning tool for beginners. The Intermediate Game could have been interesting, but little time seems to have been spent on it. It does make use of some remarkably nice counters (with some remarkably small print). Again, there is nothing new here, and players are simply pushing counters. If you're a gamer with any experience, the Advanced Game is the one that will grab your interest—and then drop it. It's one of the most splendid examples of wretched excess ever published. All that tremendous order-of-battle work (and Mick Uhl's research *is* topnotch; the order of battle here is the definitive one for wargamers) is straitjacketed by a playing board that makes no sense. The scale is totally wrong for a regimental simulation, and the game map, which is pretty but hard to read, is virtually ignored in terms of terrain effects and movement. Many of the rules are overly complex and overwrought, and the stacks of units squeezed into the small hexes could drive you crazy. What it does have is a perfectly marvelous movement system and some rules that cry out for use elsewhere. In some respects it may be a better historical tool than SPI's vaunted *Terrible Swift Sword,* but no one in his right mind would want to play it.

EVALUATION:
Presentation—Very Good
Rules—Good
Playability—Basic and Intermediate Games: Very Good; Advanced Game: Fair
Realism—Basic and Intermediate Games: Poor; Advanced Game: Very Good
Complexity—Basic Game: 2
Intermediate Game: 4
Advanced Game: 9

OVERALL EVALUATION: Fair

Lee Moves North (1973—originally *Lee At Gettysburg*)

PUBLISHER: Simulations Publications, Inc.

SUGGESTED RETAIL PRICE: $9 (plastic box)

SUBJECT: Robert E. Lee's invasions of the North in 1862 (Antietam) and 1863 (Gettysburg) during the Civil War. It covers land battle only, except for abstract naval movement for the North.

PLAYING TIME: A long evening is needed.

SCALE: This is an operational game with strategic overtones: it uses corps (North) and divisions (South). Each turn represents a week, and each hex covers about ten miles.

SIZE: The 22″ x 34″ game map uses 16mm (5/8″) hexes.

BALANCE: Very good.

KEY FEATURES: This is a simultaneous movement (SiMov) game that utilizes hidden movement by having all counters turned face down. There's considerable emphasis on supply and the problems of living off the country. The brutal Combat Results Tables are different for each side; casualties for major engagements are high. The importance of leadership (at least here) helps the South.

COMMENTS: Although the graphics are not up to the present state of the art (the map is essentially done in two colors), *Lee Moves North* is one of the best simulations of nineteenth-century operational warfare ever designed. It's also a tense, challenging game. Despised by the "hack-and-kill, I-love-a-tank" school of players, the entire thrust of the game is to win by avoiding combat. The system is based on that originally designed for SPI's *Franco-Prussian War,* but here it has been refined and honed to near perfection. While many players shrink from SiMov/hidden movement games, the problems inherent in those systems are kept to a minimum because so few counters are in use (eight or nine for each side!). If your prejudices don't prevent you from enjoying this, you may find it one of the great "sleeper" games of all time.

EVALUATION:
Presentation—Fair
Rules—Very Good
Playability—Very Good
Realism—Very Good
Complexity—6

OVERALL EVALUATION: Very Good

Objective: Atlanta (1977)

PUBLISHER: Battleline

SUGGESTED RETAIL PRICE: $11 (boxed)

SUBJECT: The Union's drive to seize Atlanta in 1865 during the Civil War.

PLAYING TIME: The shortest scenario—twelve turns—takes about six hours to complete; a full-campaign game would probably run forty-eight hours or more.

SCALE: This is an operational game with the division as the basic unit. Each strength point represents about five hundred men, and each hex covers about one and one-quarter miles.

SIZE: The 22″ x 28″ game map uses 16mm (5/8″) hexes.

BALANCE: It takes a pretty good Confederate player to win, but the heavy logistics requirements may prove too much for some Union players.

KEY FEATURES: This game places a heavy emphasis on the logistical end of the campaigning—supply lines, ammo, and all that. There is a large Movement Chart, extensive route rules, and much time spent on construction of bridges, entrenchments, and the use of slave labor. The combat-resolution system, which is fairly complex for an operational game, includes first fire, withdrawal, and assault coordination. The differential Combat Results Table has step losses and different sections for different types of terrain. The Advanced Game throws in artillery duels, railroads, and the important supply game, which is quite detailed.

COMMENTS: In terms of historical insight, at least, this is impressive. The designer—a wargame "outsider" who worked on this project for several years—has left little out. So anyone playing this will really get the feel of this difficult and little-understood campaign. Unfortunately, therein also lies the problem. *Operation: Atlanta* is too much of a simulation to be a good game, and the subject is too esoteric to hold the interests of the "historians." The Southern player is always on a depressingly worsening defensive, and that's too much of a drag to carry for more than a hundred turns. Also, because of the accuracy of the logistical requirements and the terrain problems, units move exceptionally slowly, so the game has little maneuver or fluidity. This might be accurate, but it doesn't help playability. Still, if your main interests are organizational and logistical problems, or you just plain like the Civil War, you might try it.

EVALUATION:
Presentation—Very Good
Rules—Good
Playability—Fair to Poor
Realism—Very Good
Complexity—8

OVERALL EVALUATION: Fair to Good

Road to Richmond (1977)

PUBLISHER: Simulations Publications, Inc.

SUGGESTED RETAIL PRICE: $4 (resealable plastic bag)

SUBJECT: This game simulates an engagement of the Civil War fought June 26–28, 1862, between Union troops under George B. McClellan and Confederate forces commanded by Robert E. Lee.

PLAYING TIME: The games takes one to four hours.

SCALE: This is an operational (grand-tactical) simulation with units representing brigades of infantry, cavalry, and artillery. Each hex is four hundred meters across, and each turn represents approximately two hours.

SIZE: The 17″ x 22″ map uses 19mm (3/4″) hexes.

BALANCE: This is fairly even, with a slight edge in favor of the South.

KEY FEATURES: The *Blue & Gray* system is used essentially intact with a few additional frills that include the early arrival of Union reinforcements and the characteristics of a Union staff unit. Optional rules include restrictions on advance after combat, Union siege-train artillery, and variable Confederate reinforcements and victory points.

COMMENTS: This might as well be called "Son of *Blue & Gray*." The situation presents some options for each player, and the fluidity of the battle gives it more interest than those of the QuadriGames, except for *Chickamauga*. As in those games, use of the combat-effectiveness option takes much of the sting out of attacking units, but the other options, despite some quirks, add to the realism of the simulation.

EVALUATION:
Presentation—Very Good
Rules—Good
Playability—Very Good
Realism—Good
Complexity—4

OVERALL EVALUATION: Good

Roark's Drift (1978)

PUBLISHER: Historical Alternatives

SUGGESTED RETAIL PRICE: $9.95 (boxed—barely)

SUBJECT: The game simulates the battle between Company B, 24th Welsh Regiment (British), and a huge Zulu army in Africa in the late-nineteenth century. (The same incident was depicted in the movie *Zulu*.)

PLAYING TIME: An evening is required.

SCALE: This is a close-tactical simulation: each British counter represents one man, while Zulu counters represent groups of warriors. There are ten seconds per turn and ten feet per hex.

SIZE: The 22″ x 27″ game map uses 16mm (5/8″) hexes.

BALANCE: The game favors the British, but a persistent and wily Zulu can do fairly well.

KEY FEATURES: The play sequence is Move/Move, Fire/Fire, Melee. Combat Results Table results are killed or wounded units. There are special rules for setting fire to the hospital, the use of doctors, off-board fire, and so on. There's very little new, but the application is somewhat unusual.

COMMENTS: The map doesn't fit into the box; the counters are crudely drawn; the rules have more loopholes than an insurance policy; and, what's worse, the game title misspells the site of the battle (which should be "Rorke's Drift"). Despite this, the game is a lot of fun to play. The Zulu charges are quite tense, and a careless British player can get cut to ribbons. Although lacking true historicity, it simulates the feel of the "siege" nicely. It's an interesting novelty, but it could have been done better.

EVALUATION:
Presentation—Fair
Rules—Poor
Playability—Very Good
Realism—Good
Complexity—4

OVERALL EVALUATION: Fair

Sevastopol (1978)

PUBLISHER: Simulations Publications, Inc.

SUGGESTED RETAIL PRICE: $9 (resealable

plastic bag; also available as part of *The Art of Siege* for $27, boxed)

SUBJECT: The siege of the Russian city during the Crimean War, 1854–1855. It covers land operations only; naval elements are abstracted.

PLAYING TIME: The game takes about seven hours.

SCALE: This operational game uses regiments, battalions, and batteries. Each turn represents two weeks, and the map scale is 1″ = 150 yards.

SIZE: The 22″ x 34″ game map has no hexes or areas. It's simply a map of Sevastopol.

BALANCE: Balance is acceptable, although, as in a good many sieges, the attacker seems to have a slight edge.

KEY FEATURES: See review of *Lille* in Chapter 8.

COMMENTS: Except for being somewhat less interesting than *Lille*, the game is quite similar, and most comments for that game apply here. There is, however, a bit more movement by the defenders in this case. The map is a replica of a period map of the city and is quite interesting if a bit garish.

EVALUATION:
- Presentation—Very Good
- Rules—Good
- Playability—Good
- Realism—Very Good
- Complexity—7

OVERALL EVALUATION: Good

Shenandoah (1975)

PUBLISHER: Battleline Publications

SUGGESTED RETAIL PRICE: $10 (boxed)

SUBJECT: The Civil War campaigns conducted for control of the Shenandoah Valley in 1862 and 1864.

PLAYING TIME: Short scenarios take about two to four hours. Each campaign scenario could take many, many hours.

SCALE: This operational game uses units of various sizes from brigade to divisional strength. One turn equals one day, and each hex covers two and one-half miles.

SIZE: The 22″ x 27″ game map uses 16mm (5/8″) hexes.

BALANCE: Balance is quite good and changes as each side brings in and withdraws units. Much of the balance is determined by what the players want to commit.

KEY FEATURES: The overall system for this game is unique, although individually nothing is drastically radical. Movement depends on formation and (especially) supply. In addition, units may use their movement factors to attack, move on, and attack again. The combat procedure is fairly complex: each side attacks the other in a series of rounds, and reserves are thrown in for good measure. The Combat Results Table itself is a bit convoluted: the final attack ratio is dependent on the number of hexes the player is attacking from, the terrain, and other considerations. Other rules cover weather, wagons and supply, hidden movement, partisans, destruction, devastation of hexes, garrisons, chain of command, cavalry raids, and so on. There are a large number of small scenarios—as small as one turn.

COMMENTS: This game was way ahead of its time and was poorly appreciated when released. Since it's top-heavy with unusual rules, the system is more than a little inelegant. If you have a special interest in this area, and if you can overlook the overly convoluted combat system and the opacity of some of the rules, you may appreciate some of the interesting provisions here. *Shenandoah* has a wonderful feel for the quasi-guerilla warfare that characterized the valley campaigns. Surprise is a key factor, but this is not just a combat game; organization, planning, and an iron nerve are all vital. This is one of the best operational simulations on this era available—but it's decidedly not for the timid.

EVALUATION:
- Presentation—Good
- Rules—Fair
- Playability—Fair
- Realism—Very Good
- Complexity—8

OVERALL EVALUATION: Good—but not for everyone

Source of the Nile (1978)

PUBLISHER: Discovery Games

SUGGESTED RETAIL PRICE: $11 (boxed)

SUBJECT: The game simulates the exploration of Africa at the height of European colonialism in the late-nineteenth century.

PLAYING TIME: About two hours for each expedition. The length of a total game depends on how many expeditions the players want to undertake.

SCALE: The scale here is fairly abstract, but each turn represents about a week and each hex on the map about one hundred miles.

SIZE: The 17″ x 23″ vinyl map uses 19mm (¾″) hexes.

BALANCE: Balance is something of an extraneous item here, as each player is essentially playing against himself.

KEY FEATURES: This is a unique system: the map is blank, at least to start. Players fill in the terrain as it is "discovered," and it is *not* the same from game to game. Event cards are used to randomize disasters and other relevant events. The emphasis is on planning expeditions and then sticking to the plan. There is a nice method for determining the size and location of major landmarks.

COMMENTS: This is not strictly a wargame: there are no battles. However, it is a historical simulation and one of the best in years. Each game is different, and the feel of the game depends wholly on the players, who battle not the other players but themselves and the elements. It's a great deal of fun to play, and as a bonus it provides plenty of insight into a fascinating era. At Origins '79 it was something of a surprise winner of a Charles Roberts Award.

EVALUATION:
- Presentation—Good
- Rules—Good
- Playability—Good
- Realism—Very Good
- Complexity—7

OVERALL EVALUATION: Very Good

Stonewall (1978)

PUBLISHER: Simulations Publications, Inc.

SUGGESTED RETAIL PRICE: $10 (plastic box)

SUBJECT: The Civil War Battle of Kernstown in March, 1862.

PLAYING TIME: The game takes four to five hours.

SCALE: This is a tactical game using regiments and batteries. Each turn represents twenty minutes. Each hex equals 125 yards.

SIZE: The 21″ x 33″ game map uses 16mm (⅝″) hexes.

BALANCE: Despite the great difference in overall strength between the two sides, this game has good balance—a testimony to the system's ability to depict actual, but sometimes intangible, factors.

KEY FEATURES: *Stonewall* is based on the award-winning Civil War battle system pioneered in *Terrible Swift Sword.* The emphasis is on ranged fire, command control, morale, and the integration of all these factors. Each unit has a specific type of weapon and ratings for strength and morale. Commanders are rated on their ability to control, lead, and rally units. Other rules cover ammunition, artillery accuracy, melee (which takes place within the hex rather than in adjacent hexes), gun crews, and so on. The game is quite similar to *Terrible Swift Sword,* except that gun crews have been added and the Combat Results Table for firing has been adjusted somewhat.

COMMENTS: *Stonewall,* which originally appeared in *Strategy & Tactics* magazine, was one of the most popular "issue" games in two years—not surprising in view of the system on which it's based. The *Terrible Swift Sword* system enables players to simulate Civil War battles accurately but very smoothly, and *Stonewall* has none of the bulkiness of the parent game. The number of units is small —less than fifty in total—and play proceeds rapidly. The effective complexity is not so great as it might appear; the flow of play is so natural that you rarely refer to the rules. The battle itself is fairly interesting and presents both players with many strategic and tactical problems; however, Gettysburg it is not. While too complex for the novice, this is an excellent game for introducing the "journeyman" gamer to musket-era tactical games, and it's a good evening's fun.

EVALUATION:
Presentation—Very Good
Rules—Very Good
Playability—Good
Realism—Very Good
Complexity—7

OVERALL EVALUATION: Very Good

Terrible Swift Sword (1976)

PUBLISHER: Simulations Publications, Inc.

SUGGESTED RETAIL PRICE: $22 (plastic box)

SUBJECT: The Battle of Gettysburg, July 1–3, 1863.

PLAYING TIME: It takes fifteen to twenty hours for a day's scenario; at least forty hours for the complete game.

SCALE: This is a tactical (grand-tactical) simulation. Units represent regiments of infantry and cavalry, and batteries of artillery. Each hex represents 120 yards; each turn twenty minutes.

SIZE: Three 22″ x 34″ map sheets with 16mm (⅝″) hexes overlap to form a 44″ x 56″ playing surface.

BALANCE: Four scenarios are given, as well as the campaign game. First-day action favors the Confederate forces, but other scenarios favor the Union to an increasing degree. In the campaign game chances are about even.

KEY FEATURES: *Terrible Swift Sword* introduced a new and very detailed system for simulating Civil War battles at the tactical level. Defensive fire is integrated into the attacking player's turn, which creates a more realistic flow of play. Fire combat includes provisions for enfilade fire and attacks on moving, adjacent units. Melee uses a differential Combat Results Table,

SPI's Civil War battle game Terrible Swift Sword. (PHOTO: SIMULATIONS PUBLICATIONS, INC.)

a system of losses by step reduction. This requires a second counter for every unit and can make for some large stacks. Prisoners may be taken. Formations are important for utilizing road movement and for concentrating reserves. Morale is important, and the swarm of leader counters are required for movement and rallying units. Succession of command is detailed. Other rules cover ammunition supplies, cavalry charges, night action, and construction of breastworks. There are optional rules for brigade combat effectiveness, artillery accuracy, and more detailed supply rules.

COMMENTS: *Terrible Swift Sword* provides a fascinating study of the Battle of Gettysburg. The game is immense, but so is the enjoyment. The rules for enfilade fire entice players to strive for the same tactical objectives that were important during the actual battle. The historical situation provides many opportunities for both players to exercise their initiative in bold strokes of tactical brilliance. The wealth of detail, variety of weapon types, necessity of command control, even the attractive map sheets add to the air of vivid reality. And it's more playable than you would expect of a monster this size. *Terrible Swift*

Sword won the Charles Roberts Award as the best tactical wargame of 1976.

EVALUATION:

Presentation—Very Good

Rules—Good

Playability—Fair

Realism—Very Good

Complexity—8

OVERALL EVALUATION: Very Good

CHAPTER 11

The Lights of Europe Go Out

World War I was one of mankind's greatest mistakes. It caused at least ten million deaths and the wounding or maiming of twice that number. The genetic consequences of the loss of the "best and brightest" of a generation are incalculable.

The roots of World War I lay in the rivalry between France and Germany and can be traced back through the Franco-Prussian War to the campaigns of Napoleon, when the French opposed the Austrians and Prussians. Out of the Franco-Prussian War came the unification of Germany, the acquisition of the formerly French territories of Alsace and Lorraine, and an intensification of ill will between the two countries. Otto von Bismarck, the "Iron Chancellor," saw no reason to stop there. He carefully tutored Wilhelm II during the boy's formative years and succeeded in imparting to him a vision of Germany as the leading power of Europe. Their joint ambitions could only be satisfied by another war of conquest.

As Kaiser, Wilhelm gave strong support to the German military. Simultaneously, the Krupps and other industrial barons were bringing economic and technical primacy to Germany. The leaders of the German General Staff, recognizing the possibilities inherent in the industrial complex, made plans to exercise this new strength against their old enemies.

Germany's Schlieffen Plan called for a strong "plunge" through Belgium and the Low Countries to avoid the area in south-central France that had been fortified in the years since the last war. These regions would be enveloped and the French armies divided into two portions, easily isolated and defeated in detail. The officers of the German General Staff were mostly veterans of the Franco-Prussian War, and what they intended was a quick repeat of that war—a blitzkrieg of the sort that opened World War II. No one wanted a long, drawn-out conflict. The aim was to crush the French before the Russians, who were a century behind the state-of-the-art in warfare, could mobilize and come to the aid of their allies. All the Prussians lacked was an excuse.

Politics in Europe at the time were highly convoluted. The Triple Alliance (Germany, Austria-Hungary, and Italy) and the Triple Entente (Britain, France, and Russia) represented opposite ends in European politics, but they were also beset by internal conflicts of interest. Individually and collectively, they were involved in the affairs of the Balkan states—particularly Serbia, Rumania, and Montenegro—which, along with Austria-Hungary, were also entangled in various ways with the Ottoman Empire (Turkey). Generally, Turkey was aligned with the Central Powers (the Triple Alliance) as a balance against Russia, but French economic interests in Turkey gave France strong reasons to worry about Turkey's military and political stability. While its military mechanics are wholly unrealistic, the atmosphere of complex interaction, negotiation, cross-connected intrigues, multiple alliances, and back-stabbing maneuvers of the game of *Diplomacy* does convey a fairly accurate portrait of the spirit of the times.

The excuse for war came on the morning of June 28, 1914, when Serbian nationalists assassinated the Austrian Grand Duke Franz Ferdinand, a close personal friend of Kaiser Wilhelm and heir to the throne of Austria-Hungary. In the following thirty-eight days, the political status quo disintegrated. On August 4, Austria-Hungary invaded Serbia, and Germany began to march through Belgium to France. Since Russia had committed itself to support Serbia, Wilhelm hoped the distraction would secure German territory from a Russian advance.

To German consternation, however, the British decided to honor a seventy-five-year-old commitment to guarantee Belgium neutrality. Britain dispatched a small force of 150,000 men to the continent, and used its vast naval superiority to bottle up the German fleet in the North Sea.

Nonetheless, the Germans advanced swiftly through Belgium and Luxembourg, and, despite the failure of the German high command to implement the Schlieffen Plan successfully, the entire French front was in grave jeopardy by the end of August. After fighting delaying actions along the Meuse River against the German right wing, the French and the British Expeditionary Force, by early September, had fallen back to the Marne, the last geographical obstacle between the invaders and Paris.

During the first week of September, with the French and British rapidly withdrawing before the German juggernaut, events and coincidence conspired to produce the "Miracle of the Marne." Because of the losses to his right flank, Helmuth von Moltke was required to contract the length of the German front, which then stretched from Flanders to Metz. Simultaneously, the French commander, Joffre, put strong pressure on the Germans from the "hinge" to the German right wing. In a bizarre episode on September 9, a German junior officer armed with Von Moltke's authority ordered a general retreat from the thrust of the British Expeditionary Force.

German attempts to capture Channel ports in France and Belgium were frustrated in the Battle of Ypres. Anglo-French efforts to drive the Germans back toward the Rhine were frustrated by the establishment of trenches and field fortifications by the skilled German military engineers. By the next spring, more than four hundred miles of trenches stretched from the Swiss border to the sea. The mobile phase of the war in the West was over.

Soldiers is a fine game of this early phase, illustrating the various problems that brought World War I to a standstill. Prime among these was the mismatch of tactics and weaponry.

The commanders of all the European armies had been trained in the tactics that sustained Napoleon, but modified by the experience of the Franco-Prussian War. In earlier wars, cavalry was an important arm;

artillery was used to channel the movement of troops; and there were no aircraft for observation or bombardment. By 1914, the world had changed, and—even more than in the Civil War—the military mind had not assimilated the changes.

Artillery was massive in 1914. The German armies were supported by 420mm howitzers transported on railroad cars. They carried many pieces of more than 200mm. Pieces of more than 100mm were organized in companies to support the British, German, and Austrian armies. All of these weapons were utilized for bombardment, and the "smaller" guns were frequently used in direct fire on the battlefield. Their enormous killing power was hardly suspected by generals who had learned their craft with twelve-pounders.

Bolt-action rifles had also increased the rate of fire possible to infantrymen, and this increased firepower was also an unknown quantity. The automatic weapon—the machine gun—raised the problem almost exponentially. Against the interlocking fields of fire possible to entrenched machine guns, cavalry was useless. Cavalry charges and frontal assaults with infantry that were reckless in the American Civil War were suicidal in World War I. Unfortunately, the generals who ordered such debacles were not leading them, and they lived to repeat their mistakes.

Artillery bombardment of enemy positions was futile. Once dug in, men couldn't be displaced by simple explosives. Thus, the increased firepower of the infantry had also defeated the ability of infantry as a force: they couldn't go anywhere. The defense held all the cards, and there was no way to get a new deal.

This is the major problem common to games of World War I. Though we know better today, the commanders of the time were sure that the only option was to embed the front-line troops in earthworks to secure the line. Rules to force players to adopt a similar ostrich-like attitude are frustrating. Even systems that would reward a continuous front when it is faithful to the tactics of the period would, in the case of World War I, reward tactical foolishness or downright idiocy.

Since the commanders of the time thought only massive forces in prepared positions could withstand an infantry assault, a battalion was expected to hold less than a kilometer of front. When we consider the power available in machine-gun and artillery fire, it's hard to imagine who—or what—the tacticians thought they were facing.

As a consequence, both sides spent much effort to devise shock weapons and tactics to crack the opposing line, but these were generally a case of too little, too late, and genuine solutions were ignored or misused. Poison gas was ineffective because it wasn't exploited to create breakthroughs. The delayed caution of commanders was misplaced, and the opportunity passed with the initial shock. Gas masks were developed and issued, and the stasis was resumed. German *Stossgruppen* (shock groups) were small units with heavy weapons developed as tools for breaking the enemy front, but their use was delayed and insufficient. The armored fighting vehicle (shipped to the continent from Britain under the manifest "tanks") was introduced to devastating effect near the end of the war, but its potential had existed from the beginning, and the refinement of armor tactics was left for the next war. On all fronts, commanders were "content" to sit and wait for the earth to open under the enemy.

Land games of World War I have generally been plagued by the same lack of dynamics that mired the war itself. The games are usually as dull and static as the war they mimic. It shouldn't be surprising, therefore, that, uniquely among all wars and periods, land games are less popular than air or naval ones.

The war in the air was everything the war on the ground was not: active, varied, colorful, innovative, and yet chivalric. The Battle of Jutland, the only major engagement of the entire dreadnought era and one of history's largest collections of capital ships, in a strategic sense effectively sealed the fate of the German Navy. While tactically it was as irresolute and pointless as the conflict on land, it doesn't have to be when played as a game. The potential for action and movement is there. Consequently, more people play *Richthofen's War, Jutland,* or *Dreadnought* than *Tannenberg* or the boring *1914.*

Evaluations

Diplomacy (1977—a new edition of a game originally published by Games Research in 1961)

PUBLISHER: Avalon Hill Game Co.

SUGGESTED RETAIL PRICE: $15 (boxed)

SUBJECT: This is a very abstract treatment of the diplomatic and military conflicts leading up to World War I, in and around Europe, on land and sea.

PLAYING TIME: It takes five to twelve hours.

SCALE: This is a grand-strategy game. Units are abstract armies and fleets; turns are seasonal, two per year. Instead of hexes, the board is divided into geographical areas.

SIZE: The original board was 20″ x 27″; the new one is a bit smaller.

BALANCE: England or France may have the best chance, while Italy occupies the poorest position. Germany and Austria-Hungary are the most challenging. Notwithstanding all that, the results depend *far* more on the diplomatic skill of the player than any positional advantages of any particular country.

KEY FEATURES: Although there are some provisions for fewer participants, this is really a game for seven players, each of whom runs one of the major powers of Europe prior to World War I: England, France, Germany, Italy, Austria-Hungary, Russia, and Turkey. There are only two kinds of units, of equal weight: armies and fleets. There are no movement or combat factors, no Combat Results Table, and no hexes. Movement is from one area to an adjacent one. Combat is effectively a function of movement; a move (or an attempted move) is, loosely, an attack on the province (area) and/or the unit currently occupying it (if any). Conflict is democracy in action: the majority rules. A unit that could move into a province can instead support the attack of another unit (belonging to anyone); the one with the most support "wins" and occupies the province. Ties are stalemates. The short-range goal is the occupation of certain provinces called Supply Centers, which allow a player to build (or retain) units on a one-for-one basis. The long-range goal is the occupation of more than half the supply centers on the board. The mechanics are deliberately simple (if slightly more complicated than implied here), because the real emphasis of the game is entirely different. Prior to every turn there's a short period devoted to diplomacy, the heart of *Diplomacy:* negotiations with other players, alliances, promises, cajoling, lying, double-crosses, triple-crosses —anything goes. To maintain the proper suspense and tension, and to allow alliances and betrayals a chance to work, all moves are written secretly (a much simpler process than anything Simulations Publications uses for simultaneous movement) and exposed (the very word) simultaneously.

COMMENTS: This is probably the most fa-

Avalon Hill's popular game Diplomacy. (PHOTO: AVALON HILL GAME CO.)

mous wargame in existence and is certainly the one most played by the "outside world." The double-dealing required attracts some people irresistibly and repels others just as strongly, but the interaction in *Diplomacy* is the model for all other "power politics" games. The simple mechanics are militarily unrealistic (to put it mildly), but all attempts to "improve" the game by adding CRTs, combat factors, and the other paraphernalia of conventional wargames have inevitably taken something away from the subtle machinations that make the game what it is. Opening play is somewhat stereotyped—but so is that of chess. The game does, however, suffer from two major difficulties: the number of players required and the time it takes to play. While nothing new for wargamers, anyway, the latter can be helped greatly by strictly enforcing the time limit for diplomacy periods. The most common solution to both problems is postal *Diplomacy* (playing by mail), in which the writing of fictional (and often nonsensical) "press releases" brings it close to the role-playing games of Chapter 16. By mail or in person, this is a classic game in every sense of the word.

EVALUATION:
- Presentation—Very Good
- Rules—Very Good
- Playability—Very Good (Excellent, except for the time involved)
- Realism—Poor
- Complexity—4

OVERALL EVALUATION: Excellent

Jutland (1967)

PUBLISHER: Avalon Hill Game Co.

SUGGESTED RETAIL PRICE: $12 (boxed)

SUBJECT: The naval Battle of Jutland in May, 1916, between the British and German fleets.

PLAYING TIME: The game takes five to ten hours.

SCALE: This is both a grand-tactical *and* a strategic/operational simulation. Hexes on the search-board represent 36,000 yards; on the "battle-board" (generally, the floor) four feet equal 36,000 yards. Fleet movement on the search-board is in one-hour turns. On the battle-board there are six turns per hour, and triple-length counters are used to represent individual capital ships and, in the Advanced Game, groups of light cruisers or destroyers.

SIZE: Each player uses a 8″ x 10½″ paper search-board and a 11″ x 14″ mounted Task Force Board. For maneuvering ships in battle, a flat surface at least three-by-four feet in size (preferably larger) is required.

BALANCE: German chances are best in the Basic Game, but there is a definite British edge in both versions.

KEY FEATURES: *Jutland* is a mixture of board game and miniatures systems; it is the granddaddy of many more recent naval and space wargames. "Strategic" movement—the large-scale fleet maneuvering—is conducted on two separate search-boards: paper maps kept hidden from the opposing player. When fleets make contact, activities are transferred to a *large* (table tennis size) table or chunk of floor space. Ships then maneuver in formation and fire in accordance with a pair of ruler-like devices: a maneuver gauge and a rangefinder. In the Basic Game, a simple odds/ratio Combat Results Table compares the firepower of the attacking ship(s) with the protection factor (the armor, with modifications) of the defending unit; depending on the odds and the die roll, the ship is either sunk or undamaged. Damage in the Advanced Game is solely determined by firepower; each hit reduces the target ship's firepower (and attacking capabilities) for subsequent rounds. When all boxes are checked off, or if damage taken in one round exceeds a ship's protection factor, it is sunk.

COMMENTS: This game is a novelty. A sensation when it came out, it has taken its share of flak since. It does have flaws: the strategic half of the game (on the searchboards) is anticlimactic at best; the game doesn't get going until the ships are placed on the maneuvering surface. While allowing a greater scope than pure miniatures contests, it lacks much of the convenience associated with board wargames. It requires a good deal of time and room; formations are subject to disruption by pets, children, or a misplaced foot; and some people object to crawling around on their knees to play. While valid objections, they sound a bit odd coming from those who play monster games that take up as much room, have twenty times the units, and are a lot less playable. In fact, except for the space and crawling requirements, the Basic Game is about as playable as anything around, and the Advanced Game is not terribly far behind. It's fun, and the "feel" of naval warfare is unmatched by any other game of the modern period.

EVALUATION:
- Presentation—Very Good
- Rules—Very Good
- Playability—Good
- Realism—Very Good
- Complexity—5

OVERALL EVALUATION: Good

Richthofen's War (1972)

PUBLISHER: Avalon Hill Game Co.

SUGGESTED RETAIL PRICE: $12 (boxed)

SUBJECT: Aerial combat in World War I, 1914–1916.

PLAYING TIME: An hour or less is needed for most scenarios.

SCALE: This is a close-tactical simulation; each counter represents an individual aircraft.

SIZE: The 22" x 24" game board uses 16mm (5/8") hexes.

BALANCE: Eminently adjustable due to the variety of scenarios and aircraft types available.

KEY FEATURES: While simple in application, the game system is remarkable in its portrayal of the various factors involved in aerial combat during World War I. The turn sequence is straightforward—Move/Offensive Fire/Defensive Fire—but unique speed, climbing, diving, and turn capabilities make flying one type of aircraft different from any other. All combat is ranged fire; the effects on the Combat Results Table depend on the range to the target and the type of aircraft (or, more precisely, of its machine guns) attacking; there are no combat factors *per se*. Maneuver cards aid in re-creating the tactical maneuvers of the era. One of the game's most significant features is its flexibility and variety. The Basic Game can be learned quickly by novices; various advanced and optional rules add detail at the tiniest cost in playability; by combining a long sequence of scenarios, each of which can be played in a short session, the campaign game provides the detail and strategic scope of a big game with few of the disadvantages normally attending one. Scenarios are created by choosing a general type—photo reconnaissance, trench strafing, balloon busting, and so on—and then selecting aircraft from the list provided.

COMMENTS: With the addition of the maneuver cards, *Richthofen's War* has come of age. No longer do planes simply trade shots each turn until one goes down in flames. No longer does the play revolve around the superior climbing abilities of British aircraft and the German's better machine guns. French planes remain as fragile and short-lived as ever, but they now have potential in the hands of a skillful pilot. The campaign game is almost role playing, as individual pilots are identified and their kill records increase toward the prominence and security of ace status. With twenty-three missions on the schedule, conservation of forces—discretion—is important. The game has always been among the most manageable and enjoyable in the genre and, despite its complexities, can be played by the rankest novice. Yet there is detail aplenty, and the multitude of scenarios possible will keep the game fresh long after less diversified games have become stale. It has something for everyone.

EVALUATION:
Presentation—Very Good
Rules—Very Good
Playability—Excellent
Realism—Very Good
Complexity—5

OVERALL EVALUATION: Excellent

Soldiers (1972)

PUBLISHER: Simulations Publications, Inc.

SUGGESTED RETAIL PRICE: $12 (plastic box)

SUBJECT: Land combat in the opening months of World War I, 1914–1915.

PLAYING TIME: It takes two to four hours.

SCALE: This is a tactical-level game using companies, sections, and platoons. Hexes seem to be about fifty meters across, and turns represent approximately three minutes.

SIZE: The 22″ x 34″ game map uses 16mm (5/8″) hexes.

BALANCE: This is very good.

KEY FEATURES: The odds/ratio Combat Results Table is straightforward enough (units are disrupted or eliminated if affected at all), but the odds are computed against the terrain, not the defender. Units have neither defense factors nor zones of control. There are special artillery rules that reflect the use of fragmentation ammunition, and interdiction fire can be directed against a vacant hex to make it impassable. Terrain is clustered around the edges of the board, making the center quite hazardous. There are special rules for entrenchments, unit breakdown (for some units), and cavalry, which can dismount and fight as infantry.

COMMENTS: *Soldiers* is—unfortunately—little noticed now, but as the first of the close-tactical infantry games, it is the grandparent of *Sniper!*, *Squad Leader*, and even *StarSoldier*. It was a breakthrough in design and remains a good game in its own right. It nicely simulates small-unit combat without many of the rules systems once thought necessary by gamers and designers. Most of the eight-page rules folder is devoted to scenarios and a listing of units. It's also a good teaching game. Players quickly learn the importance of reserves and supporting fire, the danger of leaving a flank open, the benefits of elevated terrain, and the value of combined arms tactics. Although it simulates the early, mobile phase of the war, it makes it easy to see why the conflict degenerated into the stalemate of trench warfare. *Soldiers* is probably the best game of land combat of the period.

EVALUATION:
- Presentation—Good
- Rules—Good to Very Good
- Playability—Very Good
- Realism—Good to Very Good
- Complexity—6

OVERALL EVALUATION: Very Good

Tannenberg (1978)

PUBLISHER: Simulations Publications, Inc.

SUGGESTED RETAIL PRICE: $10 (plastic box)

SUBJECT: The opening engagements of World War I between German and Russian forces in August, 1914.

PLAYING TIME: It takes four to six hours for the campaign game.

SCALE: This is an operational-level simulation. Most units represent brigades, divisions, or corps. Hexes are eight miles across, and each turn represents two days.

SIZE: The 16″ x 22″ game map uses 16mm (5/8″) hexes.

BALANCE: The game favors the Germans to a substantial degree.

KEY FEATURES: *Tannenberg* uses essentially the same standard rules as *The Great War in the East* QuadriGame. Each side has a tactical-competence rating, which affects movement costs, command control flexibility, the ability to retreat through enemy zones of control, and the Combat Results Table used. Combat results offer the defender a wide range of options from retreat to elimination. German hidden movement is made possible by the use of substitute counters that disguise the actual disposition of forces. Innovative rules for the important factors of leaders and supply are included along with special provisions for cavalry, fortresses and trenches, rail movement, German reinforcements, the conflict between the two Russian army commanders, and a three-player game. Scenarios include historical and free-deployment versions of the campaign and two shorter scenarios that can be played in a couple of hours.

COMMENTS: This is another one of those good-simulation-but-bad-games for which SPI has been notorious. The system suits the period and puts you solidly in the trenches. The restrictions of the supply system and the limited command control lead to massive assaults rather than anything fancy or interesting like encirclements or flanking maneuvers. The Russians particularly lack flexibility; the Germans' better tactical-competence rating allows them a bit more independence. Furthermore, the options available to the defender make it difficult for the Russians to inflict casualties, and the inability of the two commanders to cooperate hampers any attempt to break the German fortified line; the Russians can't muster a sufficient concentration of force to achieve their required objectives. As a solitaire simulation for the historian, this will do nicely (once or twice), but as a contest *Tannenberg* falls far short of the mark.

EVALUATION:
- Presentation—Very Good
- Rules—Good
- Playability—Fair to Good
- Realism—Very Good
- Complexity—7

OVERALL EVALUATION: Fair to Good

To the Green Fields Beyond (1978)
PUBLISHER: Simulations Publications, Inc.
SUGGESTED RETAIL PRICE: $10 (plastic box)
SUBJECT: The land Battle of Cambrai, November 20 to December 6, 1917.
PLAYING TIME: It takes five to eight hours for the campaign game; two to three hours for the shorter scenarios.
SCALE: This is an operational treatment of the battle. Units are predominantly brigades and regiments, with some tank and cavalry battalions. Each hex represents 1,250 yards; each turn represents one day.
SIZE: The 22″ x 34″ game map uses 19mm (¾″) hexes.
BALANCE: The Germans seem to have a slight edge, but it's not enough to bother anyone.
KEY FEATURES: The turn sequence is unusual: Artillery Bombardment/Combat/Movement/*repeat*. Tanks and *Stossgruppen* may overrun enemy units during movement and get to use the mobile Combat Results Table, which is more likely to force the enemy to retreat. Step reduction of units is the normal result of the standard CRT. Extensive rules are devoted to artillery, which is capable of no less than five distinct fire missions. Other rules cover supply, breakdown and repair of tanks, construction and demolition of bridges, command control confusion, a mandated British continual line, and various options.
COMMENTS: There is a wealth of detail here. The artillery rules, appropriately, are marvelous, and the tanks are at once a source of elation and frustration for the British player. The situation is a good one, with both players getting a chance to attack and defend. The sheer number of counters predictably slows the game, but play moves along fairly steadily except for periodic flurries of die-rolling. The map is beautifully executed, and the British units are printed in dayglow orange to allow play during a blackout. The game won a Charles Roberts Award at Origins '79.
EVALUATION:
Presentation—Very Good
Rules—Very Good
Playability—Fair to Good
Realism—Very Good
Complexity—8
OVERALL EVALUATION—Good to Very Good

World War I (1975)
PUBLISHER: Simulations Publications, Inc.
SUGGESTED RETAIL PRICE: $4 (plastic envelope)
SUBJECT: Land combat in World War I, 1914–1918.
PLAYING TIME: The game takes three to five hours.
SCALE: This is a strategic game. Units represent armies; turns last six months; and hexes equal seventy kilometers.
SIZE: The 17″ x 22″ game map uses 16mm (⅝″) hexes.
BALANCE: The historical scenario favors the Allies to a small degree; there is better balance in the free-deployment scenario.
KEY FEATURES: The central feature of *World War I* is combat resource points (CRPs). Each country has a CRP income that can be used to build new armies or forts or to satisfy losses. Until the CRPs start to run out, lines are static, and the war is one of attrition. Attacking is expensive in CRPs and is, on average, always more costly than defending. The differential CRT is very unusual: increasing the attacker's

margin doesn't increase the ratio of defender-to-attacker losses (as it would in most games); instead, it increases the magnitude of losses on *both* sides. The sequence of play is also unusual: both sides move before either attacks, and each may attack three times in one turn. Other special features include rail movement, sea movement, CRP-lending among countries, and German *Stossgruppen* (which can force a defender to retreat and so are more effective at gaining ground). Victory is determined by points that can be gained for territorial objectives, Russian surrender, and other considerations.

COMMENTS: *World War I* is a surprisingly enjoyable treatment of a fundamentally dull subject. The game simulates both the economic nature of the war and its lack of movement but—amazingly—remains fun to play. While there are a lot of rules and special cases in the game, the play mechanics are clean and fast-moving and the rules are individually simple. There is something to be said for a game in which you have twenty pieces and seldom lose any—as opposed to having hundreds, gobs of which regularly go to the dead pile. *World War I* rewards planning ahead and, particularly in the free-deployment game, offers a number of strategic options.

EVALUATION:
Presentation—Good
Rules—Good
Playability—Very Good
Realism—Good
Complexity—5

OVERALL EVALUATION: Good

CHAPTER 12

The Second War to End All Wars

WORLD WAR II is both the easiest and the hardest period to introduce. On the one hand, everybody knows something about the conflict. In our culture—considering comic books, television programs, John Wayne movies, and books from best-sellers to college texts—it's unavoidable. Those who couldn't tell a trireme from a mangonel or Frederick the Great from Ivan the Terrible have seen the attack on Pearl Harbor depicted a dozen times and know all too well what happened at Hiroshima—or Auschwitz. On the other hand, the number of games based on World War II is staggering. Not so long ago, there were probably more games on some aspect of World War II than on all other periods combined. Therefore, a book this size can do no more than scratch the surface.

The extreme popularity of the subject is not really surprising. The conflict is recent enough to be a "modern" war with tanks, planes, submarines, machine guns—all the things the average person associates with combat. Given the age of most gamers, however, the war isn't recent enough to be emotionally disturbing or too personally involving. It was also a more understandable conflict, one most people perceive as having a point. Many isolationists who deplored U.S. involvement in World War I see a genuine justification for stopping Hitler's Third Reich.

For game designers, World War II offers unparalleled variety and scope: they can choose land, sea, or air, or any possible combination; simulate offensives, counteroffensives, invasions, or holding actions; and use any scale from close tactical to grand strategic. Furthermore, the data base is good; the information needed for an accurate simulation is not classified or buried by the sands of time.

The differences between the two world wars was not so much the weapons involved. It was the level of integration of various combat forces and the extent to which the opposing armed forces benefited from an evolved doctrine for the separate use of the service arms. Military thinking finally caught up with technology. By the same token, various aspects of the military system—the concept of the nation in arms, the large-scale movement of troops and supplies by rail, and the aspects of full military production for a major war—had been developed in the wars since Napoleon. The qualitative difference of World War II stemmed from the degree to which these factors came together as major nations the world over were totally caught up in the conflagration.

Tactically, the combat methods of World War II teamed armored forces and airpower in a system of "lightning war" (*blitzkrieg*), which was able to defeat the primacy of the defense established in World War I. Aided by the extension of radio communication down to the small-unit level and by the development of sophisticated management techniques for logistical support, armored forces offered military leaders an unprecedented flexibility of operations.

The effort to maintain ascendancy on the battlefield also led to a technical competition among the warring powers for the biggest and best tanks, antitank guns, artillery pieces, and aircraft. As Nazi Germany demonstrated in Russia, the side with the best equipment didn't always win; on the other hand, American combat with the Japanese showed that without large amounts of modern weapons an adversary simply couldn't compete against the massive firepower resources the Allies could field.

Unlike the defensive struggles of World War I, the preferred strategic role in World War II was offensive. Attacks would be spearheaded by armored forces supported by airpower. Without command of the air over a local battlefield, successful offensive action was extremely difficult. The German tank commanders who survived the Normandy invasion in 1944 can attest to the effect of air strength on defensive operations as well.

Strategically, airpower was used against a "target system" selected by planners—typically, enemy production resources or transportation bottlenecks. Strategic airpower could also be used tactically. For example, this occurred when the Allies diverted heavy bombers for battlefield bombardment at Monte Cassino and at Normandy. But one of the more disturbing changes wrought by the war was the expansion of strategic target systems to include the civilian population, as pursued by Nazi Germany in the blitz on London or by the British Bomber Command raids during 1943 and 1944.

On the ground, the presence of large-scale mobile formations meant that armies had the ability to exploit local successes. With proper command control, such forces could run rings around a conventional defense. German victories achieved in the early blitzkrieg operations is sufficient comment on the viability of the "continuous line" brand of thinking bequeathed by the fighting of 1914–1918. Mobile units could outmaneuver static infantry and, unless the defense was deployed in great depth, encircle large elements of an opposing field army. While the average German advance across France in 1940 was no faster than that of an army of Napoleon's time, mobile units with adequate logistical support were individually capable of substantial gains (up to fifty miles a day), and their use did increase the pace and intensity of combat.

Logistics was even more of a problem in the Pacific theater. Supply dictated major elements of strategy, as in the case of Mountbatten's Southeast Asia Command (SEAC), which could not mount a major amphibious operation until the very end of the war. In Burma, the jungle made air supply essential and greatly increased the difficulty of ground movement. The fluid, mobile ground operations so common in the European theater were missing in the Pacific, and American forces develped their offensive by island-hopping.

Two distinct wars took place at sea. One was the war of the submarine, in which the merchantmen of the belligerents were prey to hostile submarines attempting to bar the passage of raw materials and military supplies. This conflict was the province of destroyers, escorts, aircraft, hunter-killer groups, operations analysts, and naval intelligence. Allied Anti-Submarine Warfare (ASW) was effective enough to sink 781 Axis submarines. In contrast, the Imperial Navy in the Pacific sank forty-nine American submarines—not enough to effectively reduce American attacks on Japanese shipping routes.

Another whole naval war in effect was the conflict of surface vessels. In a further demonstration of airpower, the aircraft carrier emerged as the major capital ship of the period, supplanting the battleship. Increasingly, cruisers were used in missions that previously had been given to the larger warships, which were expensive to produce and vulnerable to aircraft attack. The main technical development of the surface naval war was the application of electronic technology to warfare at sea. The use of radar revolutionized naval engagements, while radio intelligence provided important information for commanders. At the same time, the use of mobile refueling techniques extended the cruising range of fleets.

From Avalon Hill in 1961 came the first of what became a torrent of games on the period. *D-Day* modeled the campaign in western Europe from the Normandy invasion to the fall of Germany in May, 1945. Perhaps this game's most attractive feature was the provision of a series of "invasion areas" spanning the coast of France and the Low Countries; this allowed considerable strategic scope for attacker and defender. Although the game notes promised "mobile battles in central France" for intrepid Allied players who could successfully get their invasion ashore, the numerous cities and river lines made the game into an offensive against a massive fortified zone. Game enthusiasts cared not a whit: for the first time, historical game mechanics had been applied to an era of dynamic military action and, moreover, one with which gamers were familiar.

Naturally, *D-Day* did not attempt to simulate all the facets of World War II. It did introduce game mechanics for an amphibious invasion, but its general provisions for movement and land combat were carried over from previous games. *D-Day* formed the basis of a series of World War II games —*Stalingrad, Afrika Korps,* and so on—that were for some years the only ones available.

While this approach satisifed the vast majority of gamers, there were periodic complaints about historical inaccuracies and occasional expressions of doubt about the suitability of the same system for widely different types and sizes of operations. It was sometimes argued that gamers were getting no more than different scenarios for the same game. This unrest—slight as it was—was aided and abetted by a clever public relations crusade by Jim Dunnigan, head of Avalon Hill's rival, Simulations Publications. He—or his arguments—prevailed; SPI prospered; and the World War II games of the last half-dozen years have been a triumph of "realism" and novelty.

For serious enthusiasts with time on their hands, this has been a boon; they now have a boggling array of systems and subjects from which to choose, and they can learn far more from current simulations than they could from the games of Avalon Hill's "classic" period. Newcomers and more casual gamers, however, have been less well served. While short and uncomplicated "folio" games are occasionally introduced by SPI for the masses, more serious efforts by most major companies in this area are reserved for games of stupendous complexity and appalling length. Such simulations are no longer playable in the conventional sense. Playing monster games like *War in the East* or *Drang Nach Osten!* requires a whole club of gamers to be a practical proposition.

The pitfalls of this approach are clearly illustrated in *Atlantic Wall*, SPI's "remake" of *D-Day*. *Atlantic Wall* is a display of mismatched systems, one-sided contests, and complexity run rampant. Although the subject has been tackled many times—in *Normandy, Overlord, Breakout,* and *Pursuit, Cobra,* among others—and some have more successfully re-created the "mobile battles in central France," nothing has replaced *D-Day* as a strategic invasion game.

Almost certainly the most popular part of World War II (from a gaming point of view, of course) is the eastern front—initiated by *PanzerBlitz*, which also introduced tactical combat, a new scale to gaming. In the decade after *PanzerBlitz*'s debut in 1970, tactical (*Panzer '44, Panzer Leader*) and close-tactical (*Sniper!, Patrol, Squad Leader,* and *Cross of Iron*) games have been among the most popular items available. Eastern-front games on every scale appear almost faster than you can count: *Stalingrad* (which preceded the eastern-front wave), *Battle for Moscow, Barbarossa, The Russian Campaign, Kursk, Destruction of Army Group Center, Turning Point, Panzerkrieg, Panzergruppe Guderian, Kharkov, Drive on Stalingrad,* and many others, including *War in the East* and *Drang Nach Osten!*

Probably the second most popular campaign is the one fought in North Africa, and here again Avalon Hill's original entry, *Afrika Korps*, though counted out a dozen times, continues stubbornly to demonstrate its endurance. Its only remaining head-to-head competitor, SPI's *Panzer Armée Afrika*, enjoys a considerable edge in the ratings (at least in SPI's *Strategy & Tactics* magazine)—but is not played nearly so often. More recent contenders—*Rommel & Tunisia, Operation Crusader,* and the incredible *Campaign for North Africa*—are not even vaguely in the same weight class. They are endurance contests.

Other land games range from half a dozen versions of the Battle of the Bulge to grand-strategic games of the entire European theater. However, there are two other entirely different categories of games, those covering the war at sea and the few that deal exclusively with the war in the air.

The first of the naval games were *Bismarck* and *Midway*. The former used to be a fine introductory game but it was recently revised to suit somewhat more advanced tastes. Its replacement at the introductory level is a strategic game, *War at Sea*, which is the simplest of all naval games. *Midway*, meanwhile, was a mediocre effort and is little played today. Its replacement, in effect, is another Avalon Hill game, *Victory in the Pacific*, which is a more complicated version of *War at Sea*. Many of SPI's forays into the area—the misnamed *Fast Carriers*, the tedious *USN*, and the monster *War in the Pacific*—suffer from a rather pronounced lack of playability. Finally, there are several games treating the war under the sea: the solitaire *Wolfpack*, the schizophrenic *Up, Scope!*, and the popular *Submarine*.

The war in the air is skimpily covered. Strategic air games seem to be the exclusive domain of Lou Zocchi, who designed *Battle of Britain* and *Luftwaffe.* SPI's tactical air game, *Spitfire,* was less successful than treatments set in earlier and later conflicts, and the reigning duo is Battleline's *Air Force* and *Dauntless.*

Obviously, we can evaluate no more than a fraction of these games, which are, in turn, only a small portion of the games available. We can only suggest the enormous variety of a period that remains the most popular in wargaming.

Evaluations

Afrika Korps (1964)
PUBLISHER: Avalon Hill Game Co.
SUGGESTED RETAIL PRICE: $12 (boxed)
SUBJECT: The land conflict between British and German forces in North Africa, 1941–1942.
PLAYING TIME: The game takes four to six hours.
SCALE: This is an operational-level simulation; units represent regiments, brigades, and divisions. Each turn equals about two weeks, and each hex roughly represents ten to fifteen kilometers.
SIZE: The playing board is approximately 18″ x 44″. Hexes are 16mm (⅝″).
BALANCE: This is good.
KEY FEATURES: *Afrika Korps* is one of Avalon Hill's "elder statesmen." The standard Move/Attack sequence is used with no provision for mechanized aftermoves. If sufficient combat odds can be attained, automatic victory can serve something of the function of overrun attacks. Supply is vital. Terrain considerations are minimal, but there are provisions for limited sea movement and sieges of fortified cities.

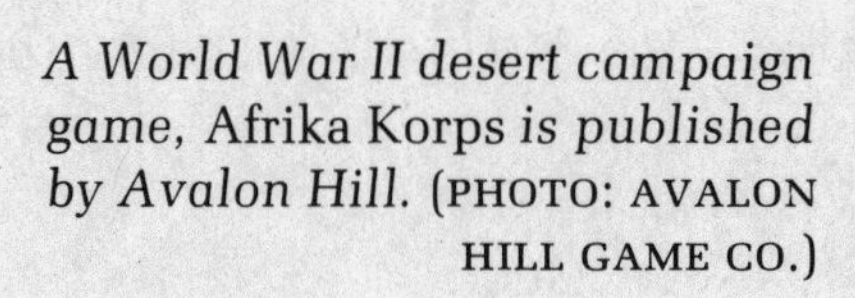
A World War II desert campaign game, Afrika Korps *is published by Avalon Hill.* (PHOTO: AVALON HILL GAME CO.)

COMMENTS: *Afrika Korps* is still one of the most frequently played games in tournament circles—a testament to its balance, durability, and playability. Perhaps too much hangs in the balance of the few die rolls for weather, which affects German reinforcement and resupply, but barring freak probabilities German chances remain good. The indirect approach, popularized in desert warfare by Field Marshal Erwin Rommel, serves equally well in the game. *Afrika Korps* is full of minor historical inaccuracies, but after more than a decade and a half, it remains the most popular game in this popular theater.

EVALUATION:
- Presentation—Fair to Good
- Rules—Good
- Playability—Very Good
- Realism—Fair to Good
- Complexity—4

OVERALL EVALUATION: Good

Air Assault on Crete and **Malta** (1978)

PUBLISHER: Avalon Hill Game Co.

SUGGESTED RETAIL PRICE: $12 (boxed)

SUBJECT: *Air Assault on Crete* simulates the Axis invasion of Crete in the spring of 1941. *Malta*, a "bonus" game, covers the hypothetical invasion of Malta. The emphasis of both is on land operations; air and sea elements are abstracted.

PLAYING TIME: Three hours or so are required for *Malta*, but a full invasion of Crete, with all optional rules, can take seven to eight hours.

SCALE: This is an operational game using battalions and companies. The scale is 1.6 kilometers per hex and eight hours per turn.

SIZE: The Crete mapboard consists of three 8" x 22" sections. *Malta* is covered on one 11" x 16" board. The Crete map is quite long—five feet, when laid out properly—so a large table is needed.

BALANCE: *Air Assault on Crete* has good balance, with both sides in the game until the end. In *Malta*, the Allied player has an advantage.

KEY FEATURES: Both games, which are packaged together, use the same system. This is fairly basic and arguably dated. The sequence is good old Move/Fight, and considering the company-level units, the Combat Results Tables are almost simplistic. Artillery is ranged. There are nice naval invasion and evacuation rules, and the airdrop procedures are handled well. Some optional rules add minor complexity, but the emphasis is on ease of play and familiarity with a tried-and-true system.

COMMENTS: This is a nice "beer-and-pretzels" game. It has none of the insight of Simulations Publications' *Descent on Crete*, but it's a heck of a lot more fun to play. Players familiar with the classic Avalon Hill systems will have little trouble with this one. Sophisticated gamers may be somewhat disappointed, but most will enjoy the challenge of planning an invasion—or trying to repel it. The box cover is grotesque.

EVALUATION:
- Presentation—Very Good (except the cover)
- Rules—Very Good
- Playability—Very Good
- Realism—Fair
- Complexity—6

OVERALL EVALUATION: Good

Avalon Hill's Air Assault on Crete *includes a "bonus game" based on the invasion of Malta.* (PHOTO: AVALON HILL GAME CO.)

Battle for Germany (1975)

PUBLISHER: Simulations Publications, Inc.

SUGGESTED RETAIL PRICE: $10 (boxed); $4 (plastic envelope)

SUBJECT: The final land offensive against Germany in 1945.

PLAYING TIME: The game takes about three hours.

SCALE: This is a strategic-operational game using corps and armies. Each hex equals sixty-seven kilometers, and each turn represents two weeks.

SIZE: The 17″ x 22″ game map uses 16mm (5/8″) hexes.

BALANCE: The Soviets will usually win, but not by a whole lot.

KEY FEATURES: The novel element is that both players play both sides (sort of). One player commands the western Allies and the eastern-front German forces; the other player commands the Soviets and the western-front Germans. The objective is to grab as much of Germany as possible. There are options for Allied-Soviet hostilities after the conquest of Germany and for three- and four-player versions of the game. Units have separate attack and (usually higher) defense factors, but the Combat Results Table favors the attacker. Attacking is voluntary, although units must stop when they first enter an opposing zone of control. There are various terrain considerations, but the system is straightforward.

COMMENTS: The split-command rules make this a novel and interesting game to play. Tactical options are somewhat limited because of the one-way direction of movement and the fortified lines on the western front. The game system, however, is smooth and playable. Although it lengthens the game, probably the most interesting version is to play out the historical game followed by the "Red Star/White Star" Soviet/Allied hostilities option.

EVALUATION:
Presentation—Good
Rules—Very Good
Playability—Very Good
Realism—Very Good
Complexity—4

OVERALL EVALUATION: Very Good

Bismarck (1962, heavily revised in 1979)

PUBLISHER: Avalon Hill Game Co.

SUGGESTED RETAIL PRICE: $12 (boxed)

SUBJECT: Sea battles between British and German naval forces, May 22 to May 28, 1941—specifically, the hunt for the German battleship *Bismarck*.

PLAYING TIME: Two to three hours are needed for the Basic Game; four to six hours if the most advanced combat system is used.

SCALE: Like *Jutland, Bismarck* combines a strategic search-board and a tactical board for resolving combat. Search-board turns equal four hours. Double-length counters represent individual capital ships or flotillas of destroyers or submarines.

SIZE: There are two 11″ x 13½″ search-boards and one 22″ x 13½″ battle-board; the game can be played on a reasonably large table. Battle-board hexes are 25mm (1″) across; sea zones on the search-board are 13mm (½″) across.

BALANCE: In the standard scenario, the British have an edge (naturally enough), but the other seven scenarios offer a range of possibilities.

KEY FEATURES: This game has a distinct split personality. The British player must search out the German vessels—by sea or air—on the search-board; then play shifts to the battle-board for combat resolution. Fog and other visibility considerations are important for the search procedure, and "shadowing" is provided for. There are two quite different combat-resolution systems. The first one, which is used in the Basic Game, is much like the original *Bismarck* system. It makes use of the hexagonal-grid battle-board but is fairly abstract. The second is the *Jutland* approach that uses a four-by-six-foot flat surface; the weapons systems are dealt with in much greater detail, and combat is less a die-rolling shootout. Other "intermediate" rules add various details and complications such as air combat, refueling, submarines, and so on.

COMMENTS: The original *Bismarck*, while distinctly unbalanced, was probably the best—and shortest—introductory wargame on the market at the time. The new version is, predictably, more detailed and more realistic, but it's not occupying the same niche. It's still a good game; it's just aimed at a broader audience. Beginners should clearly stick with the basic combat system, while more experienced players will be happier with the greater complications of the advanced version. While

Bismarck, *a naval wargame from Avalon Hill.* (PHOTO: AVALON HILL GAME CO.)

the search procedures give an authentic effect of limited intelligence, neither version is really suited for solitaire play.

EVALUATION:

Presentation—Very Good
Rules—Good
Playability—Very Good
Realism—Good (Very Good for advanced version)
Complexity—6

OVERALL EVALUATION: Very Good

Bloody Ridge (1975)

PUBLISHER: Simulations Publications, Inc.

SUGGESTED RETAIL PRICE: $4 (plastic envelope; also available as part of the *Island War* QuadriGame for $14)

SUBJECT: The Guadalcanal land campaign in September, 1942.

PLAYING TIME: Two hours are needed for either of the short scenarios; much more time is required for the full campaign.

SCALE: This is an operational game; the

basic unit is the battalion. Hexes are one thousand yards across, and turns represent two days.

SIZE: The 17″ x 22″ game map uses 16mm (⅝″) hexes.

BALANCE: Either side can win, although the Marines have an edge in the second scenario.

KEY FEATURES: *Bloody Ridge* shares a standard set of rules with the other *Island War* games. The most innovative feature of these (shared with other World War II QuadriGames) is the ranged artillery, which can make barrage attacks at long distance during the attack phase and also add to defensive strength during the defensive phase (as final protective fire). This makes employment of artillery a key game factor. Air-to-ground attacks are simulated by ground support points that work much like artillery. The *Island War* quad also features Banzai charges that double the attack factors of Japanese units but result in very heavy Japanese losses. An infiltration rule allows Japanese units limited movement through American zones of control. Victory is achieved by taking Henderson Field. Two scenarios simulate the two main Japanese offensives. For a much longer game, the players can go through the entire campaign form September 11 through November 2.

COMMENTS: *Bloody Ridge*—probably the best of the *Island War* quad—applies a very workable game system to the popular Guadalcanal campaign. The result is both a reasonable simulation of history and a playable game. By limiting the scenarios to the periods of greatest activity, the game can be played once or even twice in an afternoon.

EVALUATION:
- Presentation—Good
- Rules—Good
- Playability—Very Good
- Realism—Very Good
- Complexity—5

OVERALL EVALUATION: Good

The Campaign for North Africa (1979)

PUBLISHER: Simulations Publications, Inc.

SUGGESTED RETAIL PRICE: $33 (plastic box)

SUBJECT: The game covers all aspects—land, air, and sea—of the campaigns in North Africa, 1940–1942.

PLAYING TIME: About twenty-four hours is needed for the *shortest* scenario; the campaign game is estimated at *fifteen hundred* hours of pure playing time!

SCALE: This is an operational game. Units are battalions, individual planes, and abstracted individual ships. Each hex equals five miles, and each game turn (which is further subdivided by three) represents one week.

SIZE: The five 22″ x 34″ maps, which cover Cairo to Nofilia, use 16mm (⅝″) hexes. Since most scenarios require all five maps, players need at least a three-by-ten-foot playing surface.

BALANCE: Balance? Who cares? To *survive* is to win. (As the game emphasizes historicity, play favors the Commonwealth in most scenarios.)

KEY FEATURES: The most complex game ever designed, *The Campaign for North Africa* covers every aspect of the African campaign, from individual planes and pilots to evaporation and spillage. You name it—*The Campaign for North Africa* covers it. The design emphasis is on logistics, especially transportation and motorization. Fuel is a major factor, as are all other supplies. A novel continual movement system limits unit movements solely by how far the player wishes to push them. Combat is multileveled and intricate, with tactical overtones.

COMMENTS: *The Campaign for North Africa* is *not* a game, and to consider it as such is a big mistake. It's a history lesson—a pure simulation. On that level, it is quite an achievement; for people looking for a good "game," it is totally worthless. It's an all-consuming project that requires at least eight to ten people. It does have a remarkable order of battle (the first ever for the entire campaign on battalion level), and the system has some excellent innovations and concepts. But the game is overly complex and overlong—pure overkill. For historians and clubs only.

EVALUATION:
Presentation—Very Good
Rules—Good
Playability—Nonexistent
Realism—Excellent (at least)
Complexity—10

OVERALL EVALUATION: Very Good for historians; Very Poor for anyone else

Cross of Iron (1978)

PUBLISHER: Avalon Hill Game Co.

SUGGESTED RETAIL PRICE: $12 (boxed)

SUBJECT: Land combat on the eastern (Russo-German) front.

PLAYING TIME: The game takes to eight hours, depending on the number of units in play.

SCALE AND SIZE: In addition to one 8″ x 22″ mounted map section, there are three 8″ x 22″ paper map sections to mount over the original *Squad Leader* mapboards to make it easier to interconnect them in diverse ways. Hexes are still 19mm (¾″).

BALANCE: Very good, with some small variation among the twenty scenarios.

KEY FEATURES: *Cross of Iron* is, in effect, an expansion kit for *Squad Leader* and *cannot be played without the parent game.* It introduces a number of new units and factors into the game. A complete rewriting of the armor rules includes enough detail to please even the most devout miniatures enthusiast. New armor characteristics include individual front and flank armor modifiers, more detailed machine gun armament, special movement capabilities, target size, and range effects on penetration. Critical hits maintain the uncertainty introduced by the original fate rules. Rules covering overruns and close assaults by infantry have been rewritten, and new support weapons, terrain, specialized infantry rules, air support, snipers, and armor leaders have been added, along with even more realistic (and complicated) supplemental rules. An extensive section of questions and answers concerning peculiar situations that might arise is appended, along with a foolproof index.

COMMENTS: This is designed to be an improvement on one of the best and most successful games around. The new armor rules are far superior to the original ones, and add a depth of feeling for the subject that had been lacking. The complexities of the system are effectively summarized on the counters themselves in a masterfully done job of graphic representation. This is the last word—the state of the art

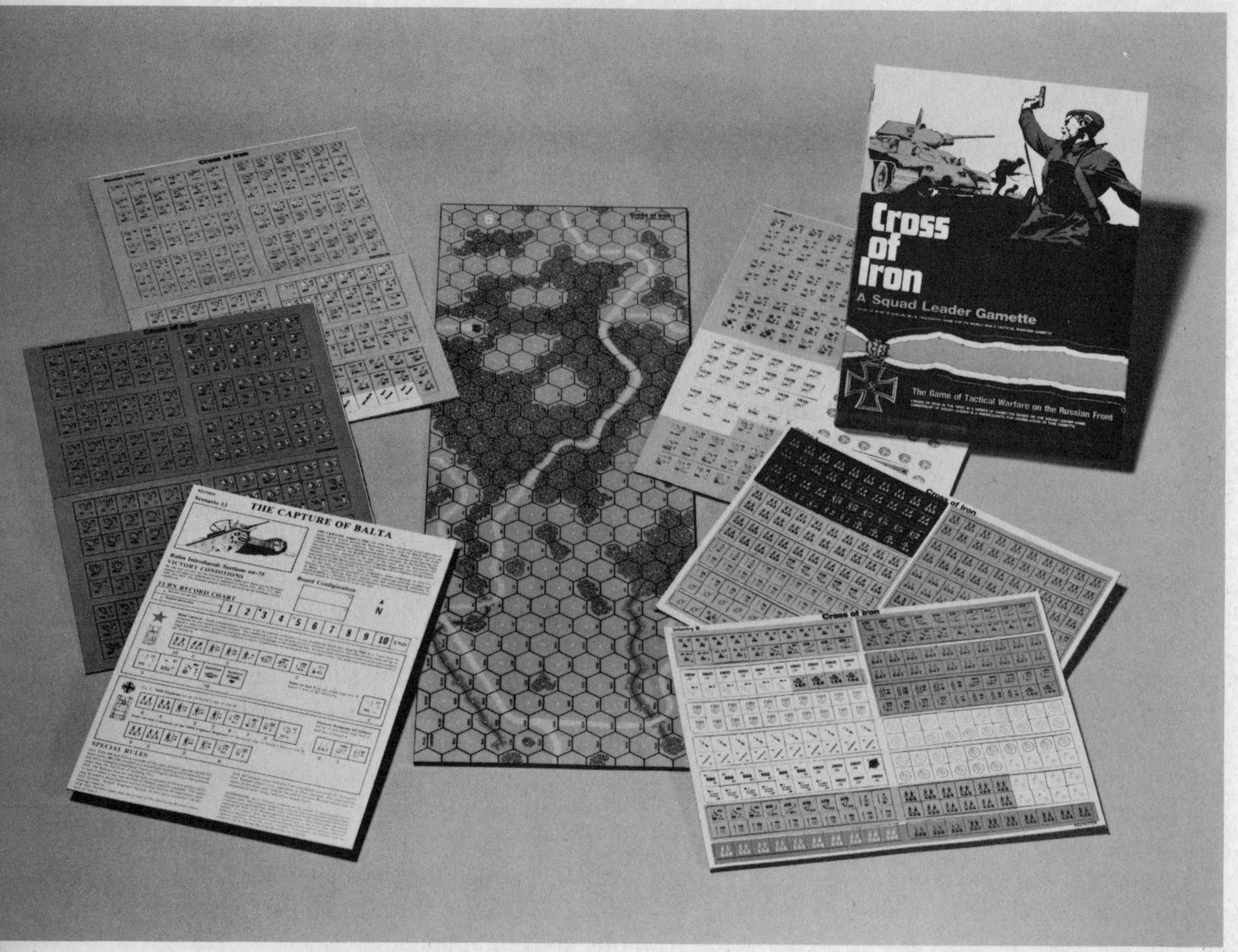

Components of Avalon Hill's wargame Cross of Iron. (PHOTO: AVALON HILL GAME CO.)

—in tactical armor games. With *Squad Leader*, it forms the most complete and realistic *playable* game system ever published. Newcomers, of course, will have to work their way up to it gradually, but they have a treat in store when they get there.

EVALUATION:
- Presentation—Excellent
- Rules—Excellent
- Playability—Good
- Realism—Excellent
- Complexity—9

OVERALL EVALUATION: Very Good (but only for the experienced)

D-Day (1961—revised in 1965 and 1977)
PUBLISHER: Avalon Hill Game Co.
SUGGESTED RETAIL PRICE: $12 (boxed)
SUBJECT: Nominally it's the invasion of Europe by the Allies in June, 1944, but actually it's the entire western-front campaign.
PLAYING TIME: The game takes three to more than ten hours.

D-Day, *an early wargame published by Avalon Hill.* (PHOTO: AVALON HILL GAME CO.)

SCALE: This is a strategic-operational game using divisions as the basic unit. Each hex equals seventeen miles, and each turn is a week.

SIZE: The 22″ x 28″ map uses 16mm (5/8″) hexes.

BALANCE: Fairly good.

KEY FEATURES: *D-Day* was the first World War II game, the first invasion game, and the first to use hexes. The turn sequence (Move/Attack), combat (with an Eliminate/Exchange/Back 2 Combat Results Table), and movement rules are from the standard Avalon Hill "classic" period, although the 1977 revision added "strategic" (double) movement for units (mostly Allied) that didn't enter opposing units' zones of control. All units had separate attack and defense factors. The supply rules changed with every revision and are now heavily dependent on Allied headquarters units. Tactical and strategic air elements are now present in abstract form.

COMMENTS: This is a genuine classic—and not just because of its historical importance. The biggest weaknesses of the original version—an abundant lack of realism, rule ambiguities, and length—were, if anything, magnified by the 1965 revision, which, among many atrocities, effectively eliminated southern France as an invasion area: an error of realism and playability. The 1977 version did a really good job of clarifying the rules, gave the HQ units something to do (as part of a far more reasonable supply system), and, through the use of Allied airpower and the strategic movement, finally allowed some of those fairly "mobile battles in central France." For a game of its complexity level, it's still quite long (though nothing like the modern three-month-long monster games)—at least if the game drags out to its full length—but the strategic possibilities continue to make it attractive.

EVALUATION:

Presentation—Fair to Good

Rules—Very Good
Playability—Good
Realism—Good (in 1977 version)
Complexity—5
OVERALL EVALUATION: Good

Drang Nach Osten! and **Unentschieden** (1973)

PUBLISHER: Game Designers' Workshop

SUGGESTED RETAIL PRICE: $14.75 for *Drang Nach Osten!*, $13.85 for *Unentschieden* (resealable plastic bags; these games are sold separately but should be purchased together)

SUBJECT: The German invasion of Russia, 1941–1945; land, air, and some sea elements are covered.

PLAYING TIME: This may not be the longest game in existence, but it's close. It could be two years of weekend playing.

SCALE: This is an operational-level game using divisions and various other units; planes are in groups of forty to sixty aircraft.

SIZE: This is monster-game territory! There are eight maps that, when placed together, will take about two pool tables to support. Plan for a *very* large playing surface plus an additional area for charts and counters.

BALANCE: Balance? The Germans roll. (Changes to the supply rules in newer editions have altered this somewhat, but their panzers are still too strong, and only a top group of Russian players will be able to overcome this and other order-of-battle shortcomings.)

KEY FEATURES: These games no longer look quite so complex as they did in 1973, when monster games were still virgin territory, but on this scale tiddledywinks would be complicated. The game sequence is simple: move, air game, land combat, exploitation (second movement/motorized). The odds/ratio Combat Results Table is fairly simple, with eliminations, exchanges, and retreats in various combinations. Stacking is handled in typically overwrought GDW fashion by assigning points to each unit type. Armor units are distinguished (at length), and engineers have lots to do. Units are supplied within the radius of a supply counter. The complex air rules virtually constitute a separate game, with a wide variety of missions and a tactically flavored combat system. Weather is accounted for and moves from map to map in nice fashion. Special rules cover movement of Russian industry, the Finns, the Arctic, first-turn surprise, airborne units, flak, railroad guns, NKVD units, HQs, partisans, and so forth. Unlike *War in the East*, there is no production schedule; everything arrives via the Reinforcement Table. There are *lots* of charts.

COMMENTS: This is a fabled game combination, and rightfully so—but not for the reason you might think. It's a legend because it proved big games—nay, monster games—had a place in the hobby. Singlehandedly, it made GDW a going concern. But as a game—and, for that matter, as a simulation—it's sucking wind uphill. Even GDW's excellent staff admits the order-of-battle work is weak compared to today's state of the art. While the game system and mechanics are quite good (as is shown by the worth of its sibling Europa games, such as *Narvik*), this is a poorly balanced game. Playability is limited to large groups with plenty of time on their hands, and the unit crush is almost unbearable. Its significance far exceeds its intrinsic merit. Saying *Drang Nach Osten!* is the best game of all time—as a magazine poll did some years ago—is like saying that *Superman* is the world's greatest movie because it cost more than any other. This game is like a brontosaurus: a fossilized relic best viewed from a safe distance.

EVALUATION:
- Presentation—Good
- Rules—Fair
- Playability—Very Poor
- Realism—Good
- Complexity—9

OVERALL EVALUATION: Not for public consumption.

Luftwaffe (1971)

PUBLISHER: Avalon Hill Game Co.

SUGGESTED RETAIL PRICE: $12 (boxed)

SUBJECT: The air war in Europe, 1943–1945.

PLAYING TIME: The game takes two to six hours, depending on the scenario.

SCALE: This could be considered a strategic-level game with tactical overtones. Units are squadrons of aircraft, and hexes are twenty miles across. Turns represent eighteen minutes but are intended to typify the operations of three months.

SIZE: The 22" x 24" game map uses 16mm (5/8") hexes.

BALANCE: The Allies have an edge, but it is not extreme.

KEY FEATURES: This game is a descendant of the early *Battle of Britain,* a Lou Zocchi design. The operations of a quarter of a year are modeled from flying one mission lasting six hours. Victory is determined by how many of the specified (and variously rated) sites across Europe are successfully bombed by the Allies. The air-to-air combat between intercepting fighters and the Allied bombers or their escort fighters is quite like simplified plane-to-plane combat. In the advanced versions are provisions for production of new aircraft types, the effect of Allied bombardment on critical consumables, and the loss and replacement of experienced pilots.

COMMENTS: *Luftwaffe* is the best game of its kind; it's also very nearly the *only* game of its kind. Subsumed by the air modules of monster games like *World War II* and *Drang Nach Osten!,* the strategic air game has all but disappeared. Aside from being somewhat unbalanced, the game's biggest flaws concern realism. The Combat Results Table is much too uncertain; the extreme results are possible but shouldn't be as likely as the more moderate outcomes. Some gamers simply disregard them, but a better method is to use two dice instead of one, juggle the figures, and rearrange the CRT so that the outcomes resemble the familiar bell-shaped curve (with the extreme results being quite unlikely). Forcing all planes of a given type to drop their tanks at once helps the playability but it's obviously unrealistic; however, it does reinforce the notion of keeping such planes together—a good tactic, anyway. Despite such quibbles, it's a playable and challenging game.

EVALUATION:
- Presentation—Good to Very Good
- Rules—Good
- Playability—Very Good
- Realism—Fair to Good
- Complexity—6

OVERALL EVALUATION: Good

PanzerBlitz (1970)
PUBLISHER: Avalon Hill Game Co.
SUGGESTED RETAIL PRICE: $12 (boxed)
SUBJECT: Land combat on the eastern front, 1941–1944.
PLAYING TIME: Two to seven hours, depending on the scenario.
SCALE: This is a tactical simulation. The basic units are companies (Russian) and platoons (German). Each hex represents 250 meters, and each turn equals six minutes.
SIZE: The three 8″ x 22″ isomorphic boards use 19mm (¾″) hexes and fit together in a variety of ways depending on the scenario.
BALANCE: Balance depends on the scenario; some early problems in this area were corrected in subsequent editions.
KEY FEATURES: *PanzerBlitz* was a watershed design. It was the first to simulate World War II events at the tactical level, the first to treat the differences between armor and infantry as more than a distinction in attack or movement factors, and the first to develop a real sequence of play, with different events occurring at different stages. It also introduced ranged combat on land —fire at a distance—and a Combat Results Table that did not affect the attacker. Different kinds of attacks—direct, indirect, close assault, and overrun—weapon ammunition, and targets made combat a much more complicated matter than before. Line-of-sight rules were necessary. Instead of the old-fashioned counter symbols, attractive silhouettes were printed on extra-large counters. Hexes were also made bigger than the previous norm, and they were on a new kind of board—actually, three "isomorphic" or "geomorphic" maps of typical terrain that could be fitted together in a variety of ways according to the requirements of the individual scenarios. There were also various optional and experimental rules, and later editions incorporated some of the innovations and modifications of its western-front sibling/descendant, *Panzer Leader*.

PanzerBlitz, *published by Avalon Hill.* (PHOTO: AVALON HILL GAME CO.)

COMMENTS: This was an enormously important game—really the first to break out of the "classic" Avalon Hill mode. Its success led directly to the popularity of the eastern front, of tactical games, and of innovative designs. It is also a very good game that is fluid in play, exciting, and colorful. It's a good model of combined arms tactics, and it leads players to use historical fighting techniques. Its biggest flaw is a product of the spotting rules, which allowed units to skulk from woods hex to woods hex without being fired on —a pattern known as the "panzerbush syndrome." This can be countered by increasing the spotting distance into such terrain. An alternative is an option in later editions: opportunity fire—defensive fire occurring during the *attacker's* movement phase—considered by some to be essential for simulations at this level. Another complaint is the high—perhaps excessive —effectiveness given high-explosive artillery by the indirect fire rules, a problem exacerbated by the optional rules allowing free spotting. Nonetheless, the units are varied and interesting, and the game remains very popular. *Panzer Leader* and *The Arab-Israeli Wars* are applications of the same system to other settings; they are attractive alternatives—but not replacements—for *PanzerBlitz*.

EVALUATION:
Presentation—Excellent
Rules—Good
Playability—Very Good
Realism—Good
Complexity—7

OVERALL EVALUATION: Very Good

Panzer '44 (1975)

PUBLISHER: Simulations Publications, Inc.

SUGGESTED RETAIL PRICE: $10 (plastic box)

SUBJECT: Land combat on the western front from June, 1944, to March, 1945.

PLAYING TIME: The game takes three to five hours.

SCALE: This is a tactical-level simulation; the basic unit is the platoon. Each hex is equivalent to two hundred meters; each turn represents up to six minutes.

SIZE: The 22″ x 34″ game map uses 16mm (5/8″) hexes.

BALANCE: Generally good, depending on the scenario.

KEY FEATURES: This is SPI's second-generation World War II tactical armor game (following *Kampfpanzer* and *Desert War*). Rather than using fully plotted simultaneous movement, *Panzer '44* requires detailed firing plots (making combat effectively simultaneous), but only a written intention to move. There are various forms of ranged fire, including direct (normal) fire, defensive opportunity fire (which adds realism but makes play by mail next to impossible), and indirect fire, which may come from off-board artillery. There are also provisions for close air support, various ammunition types, minefields, and entrenchments. Command control is instituted through a notorious system that makes use of the code number of the hexes that units occupy.

COMMENTS: Despite the typically peculiar ratings of games in *Strategy & Tactics* magazine, this is, if anything, *more* complicated than its Avalon Hill analogue, *Panzer Leader*—not less. While the system used is not as tedious as full, written orders for movement and combat, it's unsuited for anything but two-player face-to-face play, even if it's marginally more realistic than pure sequential turns. Adding a massive dose of pure chance, the absurd command control rules cannot be justified on the grounds of either realism or playability. If you must have that sort of thing, using a decimal die or chit set to check each unit's status independently will slow the game even more, but it's preferable to the old method. Despite the offensively useless infantry, *Panzer '44* isn't a bad game and when published it represented something of an advance in the state of the art of tactical armor simulations. But compared to its Avalon Hill counterparts, its attractions are few.

EVALUATION:
Presentation—Good
Rules—Good
Playability—Fair to Good
Realism—Good to Very Good
Complexity—7+

OVERALL EVALUATION: Good

Panzergruppe Guderian (1976)

PUBLISHER: Simulations Publications, Inc.

SUGGESTED RETAIL PRICE: $14 (boxed—less in other packaging)

SUBJECT: The German drive across the Dnepr River in the summer of 1941.

PLAYING TIME: The game takes four to six hours.

SCALE: This is an operational-level simulation using divisions and (for German armor units) regiments. Each hex represents 10.5 kilometers, and each turn equals two days.

SIZE: The 22″ x 32″ game board uses 16mm (5/8″) hexes.

BALANCE: The game is tilted toward a German marginal victory.

KEY FEATURES: The game is built around the idea of untried units, a system first used on a smaller scale in *Invasion: America*. Russian untried units have their combat strengths hidden until they have been involved in combat. German armor and mechanized infantry divisions display what is sometimes known as "regimental (or 'divisional,' depending on how you want to look at it) integrity." The two or three regimental counters making up a division receive combat bonuses when attacking in concert. A double-impulse movement system is employed; overruns and combat occur between the two movement phases. Soviet leaders are essential for maintaining supply. Air interdiction of transportation routes is handled in an abstract but efficient manner.

COMMENTS: *Panzergruppe Guderian* was an advance in the state of the art for operational simulations of World War II. The system of untried units is a novel and interesting way to handle the qualitative differences between the German and Russian organizations during the period. The extreme variation in actual combat strengths is, however, too great for either realism or, certainly, play balance—which is not one of this game's strong points. The situation makes the Russian player the defender, and the rules give him no chance for aggressive action. Although the system is admirably adapted for solitaire play, in a two-player game the Russian would die of frustration—if boredom didn't put him to sleep. Although not, of course, as innovative or significant, later games using the same system—*Cobra, Kharkov*, and the much longer and bigger *Drive on Stalingrad*—are better balanced and, consequently, somewhat more fun to play.

EVALUATION:
Presentation—Very Good
Rules—Very Good
Playability—Good to Very Good
Realism—Very Good
Complexity—7

OVERALL EVALUATION: solitaire—Very Good; two players—Fair to Good

Panzer Leader (1974)

PUBLISHER: Avalon Hill Game Co.

SUGGESTED RETAIL PRICE: $12 (boxed)

SUBJECT: Land combat on the western front, 1944–1945.

PLAYING TIME: Two to seven hours, depending on the scenario.

SCALE: This is a tactical simulation. The basic unit is the platoon, but individual aircraft are also represented. Hexes are 250 meters across, and each turn equals six minutes.

SIZE: Four 8″ x 22″ geomorphic boards using 19mm (¾″) hexes are included, but they're not all normally in play at once. The usual playing surface is 8″ x 44″ or 22″ x 24″ at most.

BALANCE: This is generally good, although it varies with the scenario.

KEY FEATURES: *Panzer Leader* follows in the footsteps of *PanzerBlitz*, but with four years' worth of improvements and additions. Ground-support aircraft, antiaircraft fire, and amphibious landings are new to *Panzer Leader*. Engineers have special capabilities in combat and collateral applications. Optional and experimental rules include double-time infantry

Panzer Leader, *a game of tactical warfare on the western front.* (PHOTO: AVALON HILL GAME CO.)

movement, naval gunfire support, opportunity fire, and field-of-fire considerations. Twenty scenarios are included.

COMMENTS: While the units individually don't have quite the interest of some employed on the eastern front (in *PanzerBlitz*), there's a bit more variety overall and a bit more realism in this western-front version of *PanzerBlitz*. Choosing between them is little more than personal preference in locale and battles.

EVALUATION:
- Presentation—Excellent
- Rules—Good
- Playability—Very Good
- Realism—Good to Very Good
- Complexity—7

OVERALL EVALUATION: Very Good

The Russian Campaign (1974, 1976)

PUBLISHER: Avalon Hill Game Co.

SUGGESTED RETAIL PRICE: $12 (boxed)

SUBJECT: The eastern (Russo-German) front, 1941–1945.

PLAYING TIME: Six to ten hours are needed for the full game.

SCALE: This is a strategic-level simulation. Units represent corps or armies. Each hex is equivalent to about forty kilometers; each turn represents two months.

SIZE: The 22″ x 27″ game board uses 16mm (5/8″) hexes. A separate area is needed to set up two 8½″ x 11″ order-of-battle cards.

BALANCE: Very good; the standard victory conditions heavily favor a draw. Shorter scenarios vary widely in balance.

KEY FEATURES: *The Russian Campaign* features a fairly standard double-impulse movement system with one important twist: there's a second combat phase after the second movement phase. Weather effects are very important for movement costs, especially during the second movement impulse. Combat is odds/ratio with some unusual results possible. There are additional rules for abstract rail and sea movement, rail supply, partisans, paratroops, and reinforcements. Also included are an appendix to help the beginner comprehend the railroad rules and a number of scenario variations.

COMMENTS: *The Russian Campaign* is often referred to as an improved *Stalingrad*. It's not really a simulation of World War II unless the players want it to be. The victory conditions reflect the real objectives of the respective combatants, but the methods of attaining those objectives are left to the player. The systems in the game abstract most effects to a vast extent, but the end result, the feel for those effects, is most authentic. The Germans sweep through Russia during the early summers and cling desperately to their gains through the winters. If you desire a simulation from which to study the unfolding of the invasion of Russia, *The Russian Campaign* is not it. But if you are looking for an effective and challenging reproduction of the campaign, with a multitude of options integrated into an extremely playable system for both face-to-face and solitaire gaming, you need look no further.

EVALUATION:
- Presentation—Good
- Rules—Good
- Playability—Very Good (except for time involved)
- Realism—Good
- Complexity—6

OVERALL EVALUATION: Very Good

Sniper! (1973)

PUBLISHER: Simulations Publications, Inc.

SUGGESTED RETAIL PRICE: $14 (boxed—less in other packaging)

SUBJECT: Land combat (small-unit actions) in Europe, 1941–1945.

PLAYING TIME: Two to four hours are required.

SCALE: This is a close-tactical simulation; the units are individual men. Each turn equals thirty seconds; each hex represents approximately ten feet.

SIZE: The 22″ x 34″ map uses 16mm (5/8″) hexes.

BALANCE: Varies according to the scenario.

KEY FEATURES: This is the smallest-scale game available. It uses plotted simultaneous movement (SiMov). During the plot phase, the actions of all men are determined. After fire is resolved, units move; opportunity fire occurs; and projectiles detonate. Both panic (command control) and preservation rules are included; the former is based on the hex numbers, and the latter serves, among other functions, as a check on game length. Various types of weapons and explosives are provided. Tanks, trucks, and half-tracks occupy six hexes apiece. Buildings are represented somewhat abstractly in three dimensions. Options include the use of dummy counters and hidden movement.

COMMENTS: *Sniper!* was the first man-to-man simulation to hit the shelves and as such suffers from being outdated by more recent designs. (Of course, it was somewhat ahead of its time to begin with.) The cost in playability—especially solitaire playability—of this brand of SiMov is high. The sighting rules are poor, and players would do better simply to precede the plot phase with a separate sighting phase. The unrealistic command control system is justifiably notorious and should be disregarded or replaced with another —any other—alternative. Much heralded at its appearance, *Sniper!* shows its age. It offers variety and a wealth of fine detail for those who find the subject and scale irresistible, but most gamers will find it too cumbersome. Many of its features and flavor are shared with *Patrol* (an "outdoor" wilderness version) and *Star-Soldier* (a science-fiction game).

EVALUATION:

Presentation—Good
Rules—Good
Playability—Fair to Good
Realism—Good
Complexity—8

OVERALL EVALUATION: Fair

Squad Leader (1977)

PUBLISHER: Avalon Hill Game Co.

SUGGESTED RETAIL PRICE: $12 (boxed)

SUBJECT: Land combat in World War II on a variety of fronts.

PLAYING TIME: Two to seven hours are needed, depending on the scenario.

SCALE: This is a close-tactical game on the squad level, with individual leaders and heavy weapons.

SIZE: Four 8″ x 22″ geomorphic mapboards have 19mm (3/4″) hexes. The arrangements depend on the scenario, but most can be played on a standard-sized table.

BALANCE: Fairly good in most scenarios.

KEY FEATURES: This is a highly innovative game system with several "new" concepts. Although many of the basic rules are familiar, the sequence of play—while bearing a family resemblance to other tactical simulations—is complexly interwoven and quite long. Special rules cover carrying heavy weapons, a variety of morale types, fate, flamethrowers, armored vehicles, radio, running in and out of buildings, and so on. There are several different Combat Results Tables covering the wide variety of combat capabilities, and there is a great deal of terrain represented on a beautiful game map. The rules are presented in the Avalon Hill applied-learning approach: each scenario adds new rules. There are several spinoffs and additions (released and planned): for example, *Cross of Iron*.

COMMENTS: *Squad Leader* was a Charles Roberts Award winner for 1977—and deservedly so. It is probably the most popular tactical World War II game since *PanzerBlitz*. The game is exciting, colorful, and almost endless in its variations and scenario possibilities. While the play sequence (complicated as it is) is geared more to fun than to an accurate representation of a squad-level firefight, the game does give the players a remarkable feel for close-tactical combat. Two other factors contribute to its success: the game has been given a topnotch physical presentation by Avalon Hill, and the charts have been kept to a minimum. This allows the players to enjoy the game without resorting to charts and rules at every step. Although clearly intended only for advanced players, *Squad Leader* is not unplayably long and does reward the time spent learning the rules. It seems to be that rare bird: an instant success with staying power.

EVALUATION:
Presentation—Excellent
Rules—Very Good
Playability—Good to Very Good
Realism—Very Good
Complexity—8

OVERALL EVALUATION: Very Good

Stalingrad (1963)

PUBLISHER: Avalon Hill Game Co.

SUGGESTED RETAIL PRICE: $12 (boxed)

SUBJECT: The eastern-front (Russo-German) land campaign, 1941–1943.

PLAYING TIME: Four to eight hours are required.

SCALE: This is a strategic-operational game using corps (Germany) and armies (Russia). Hexes are thirty-five miles across; each turn represents one month.

SIZE: The 22″ x 28″ board uses 16mm (5/8″) hexes.

BALANCE: In the standard version, the Russians almost always win, although the Germans are usually considered the more interesting side to play.

KEY FEATURES: *Stalingrad* uses the basic, original Avalon Hill (*D-Day*) system with a few additions: railroads, a weather table, and replacements dependent on geographic areas.

COMMENTS: *Stalingrad* is another classic and was for many years one of the most popular wargames around. It is still very common at conventions and other face-to-face meetings. Play balance is an obvious weakness: since the original made it almost impossible for the Germans to win, innumerable variants sprang up over the years to even things up. The other standard objection is on historical grounds: the game bears little resemblance to the actual campaign. Nonetheless, it was state of the art when first

Stalingrad, *a World War II game based on the campaign in Russia from 1941 to 1945.* (PHOTO: AVALON HILL GAME CO.)

produced, and it remains a playable and enjoyable game.

EVALUATION:
- Presentation—Good
- Rules—Good
- Playability—Good to Very Good
- Realism—Poor
- Complexity—4

OVERALL EVALUATION: Good

Submarine (1978—a new edition of a game originally published by Battleline in 1976)

PUBLISHER: Avalon Hill Game Co.

SUGGESTED RETAIL PRICE: $12 (boxed)

SUBJECT: Submarine warfare during World War II.

PLAYING TIME: One to three hours are needed per scenario; more than forty hours for the full campaign.

SCALE: This is a tactical-level simulation; double-length counters represent individual ships or submarines.

SIZE: Three 11″ x 27″ geomorphic boards employing 13mm (½″) hexes form a 33″ x 27″ playing surface. However, since they may shift along with the course of the battle, a somewhat larger area is preferred.

BALANCE: Reasonable, although the convoys have a slight edge.

KEY FEATURES: The heart of the game is the hidden-submarine movement and sonar-search procedure. Logs must be maintained for each submarine, with all movement, depth charges, and firing recorded. Convoy movement is plotted three turns in advance, while escort plots consist only of the present speed of the vessel. The rigid requirements of convoys are one of the few advantages of the harried submarine commander. Torpedo fire is han-

Submarine, *an Avalon Hill game that uses a featureless board.* (PHOTO: AVALON HILL GAME CO.)

dled simply, and reloading of torpedo tubes is a primary consideration. Submarine detection gear (radar, sonar, and so on) and antisubmarine weaponry vary according to the time frame of the scenarios. The Advanced Game adds considerably more complicated options.

COMMENTS: The hidden-submarine movement so essential to the game is handled well enough that the overall playability remains high, although the loss in solitaire playability is more significant. Cross-referencing the variety of weapons systems with the time period of the scenarios gives an interesting slant on the effects of technological advancement on warfare. The dangers of commanding a submarine are well presented, as are the difficulties of the escort commander in locating the attackers in order to employ countermeasures. Perhaps the biggest flaw is that there is an aura of sameness from scenario to scenario; different weapon types and situations are not quite sufficient to keep play truly diversified over a long period.

EVALUATION:

Presentation—Very Good
Rules—Good to Very Good
Playability—Good to Very Good
Realism—Good to Very Good
Complexity—6

OVERALL EVALUATION: Good

Third Reich (1976)

PUBLISHER: Avalon Hill Game Co.

SUGGESTED RETAIL PRICE: $12 (boxed)

SUBJECT: The decline and fall of the Axis powers in Europe, including land, air, and sea aspects.

PLAYING TIME: Six to twelve hours are needed, depending on the scenario.

SCALE: This is a grand-strategic simulation using armies, corps, fleets, and air forces. Each turn equals three months, and each hex is roughly equivalent to fifty kilometers.

SIZE: Four 8″ x 22″ boards employing 16mm (⅝″) hexes form a 22″ x 32″ playing surface. Some additional space is required for the scenario cards.

BALANCE: This is good.

KEY FEATURES: The central feature of an unusual system is the Basic Resource Points (BRPs), which represent the economic factors in modern warfare. These are spent for a number of purposes: many active-front options, declarations of war, building or rebuilding combat units, and so forth. Conquest of minor countries is essential for enlarging the BRP base of a nation, and extensive rules detailing major-country intervention and secondary conquests are presented along with provisions for BRP lend-lease. Other rules cover bridgeheads, amphibious assaults, sea and air missions, and convoys. There are three shorter scenarios in addition to the full campaign game plus provisions for multiplayer alliances.

COMMENTS: The original and innovative

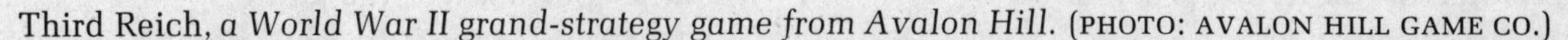
Third Reich, *a World War II grand-strategy game from Avalon Hill.* (PHOTO: AVALON HILL GAME CO.)

game system of *Third Reich* is a refreshing change from the sometimes hackneyed design features of many modern wargames. Players are given a wealth of strategic options; yet the events of the war are re-created by making the historical objectives desirable in game terms. All the major elements of the war are here, but the wealth of detail never overwhelms the game (as it does in the typical "monster" game). *Third Reich* is a Charles Roberts Award winner; it is easily the finest strategic-level simulation of the war.

EVALUATION:
Presentation—Very Good
Rules—Very Good
Playability—Good
Realism—Very Good
Complexity—8

OVERALL EVALUATION: Very Good

Unentschieden (See *Drang Nach Osten!*)

Victory in the Pacific (1977)

PUBLISHER: Avalon Hill Game Co.

SUGGESTED RETAIL PRICE: $12 (boxed)

SUBJECT: The naval war (including air aspects) in the Pacific.

PLAYING TIME: Eight hours are needed for a complete game.

SCALE: This is a strategic game using large areas rather than hexes. Turns are about three months in length, while units are individual ships, land-based air forces, and Marine divisions.

SIZE: There are two 14" x 22" boards that fit together, but considerable additional space is required. A large table is preferred.

BALANCE: Debatable and dependent on the experience of the players involved; the overall edge is to the American forces.

KEY FEATURES: This is essentially the *War at Sea* system with elaborations. Each player places patrolling ships, raiding ships, land-based air units, and amphibious units in a fixed sequence (the American moving last). After all movement is completed, ships and air units in each of the sea areas conduct attacks on enemy units in the area until one side withdraws or is destroyed. Possession of a limited number of bases in the sea areas allows a player to use his land-based air assets, so strategies generally involve the capturing of bases to extend a player's area of influence. Although the Japanese player always loses the war, he may, because of the victory points, win the game.

COMMENTS: *Victory in the Pacific* is an enjoyable game to play—partly because of the very pleasing, high-quality game components, and partly because the game mechanics work so well. Although hardly realistic, it is certainly challenging. The game is highly amenable to analysis and has been subjected to a lot of the same. Some people claim they cannot lose if they attack Pearl Harbor and attempt to take it. In our opinion the Japanese, unless extremely lucky, will usually lose—but not by much—and this slight inequity could be corrected by giving the Japanese an extra victory point or two per turn. Despite its abstract treatment, the game is quite popular (if not with the readers of *Strategy & Tactics* magazine) and won a Charles Roberts Award.

EVALUATION:
Presentation—Excellent
Rules—Fair to Good

Victory in the Pacific, *published by Avalon Hill.* (PHOTO: AVALON HILL GAME CO.)

Playability—Very Good
Realism—Poor
Complexity—5

OVERALL EVALUATION: Very Good

Wacht am Rhein (1977)

PUBLISHER: Simulations Publications, Inc.

SUGGESTED RETAIL PRICE: $24 (plastic box)

SUBJECT: The Battle of the Bulge, December 16, 1944, to January 2, 1945.

PLAYING TIME: Ten to fifteen hours are needed for the smaller scenarios; more than forty hours for the campaign game.

SCALE: This is a grand-tactical simulation; units are battalions and companies. Each hex is a mile across, and one daylight turn equals 4.5 hours.

SIZE: Four 22″ x 34″ game maps—using 16mm (5/8″) hexes—overlap to form a 44″ x 68″ playing surface. Additional space is needed to set up the two 11″ x 17″ turn-record and reinforcement charts.

BALANCE: The Allies have an edge, but it is smaller in the shorter scenarios.

KEY FEATURES: A double-impulse movement system employs both field and travel modes. Bridge-blowing and -building (and therefore engineers) play a vital part in movement and also determine supply paths. Stacking varies by mode and unit type, and the combined arms rules make precise and varied stacking essential. Armor, reconnaissance, artillery, and headquarters units are extensively differentiated. Air power is abstracted by means of air points allocated to various missions. There are provisions for weather and regimental integrity. Optional rules introduce further complexities.

COMMENTS: *Wacht am Rhein* is perhaps the ultimate simulation of the Battle of the Bulge, long a favorite gaming topic. The wealth of detail reduces the playability of the game, but you don't purchase a game of this physical immensity for playability anyway, and the illusion of reality created is intense. The emphasis is on the interaction between the many types of units portrayed in the game, and the various bonuses accorded cause players to strive for real tactical ideals. The optional rules add detail but cannot improve much on the basic system. For true Bulge enthusiasts, it will provide endless (well, forty-plus) hours of engrossing and educational conflict.

EVALUATION:

Presentation—Very Good
Rules—Very Good
Playability—Fair (at best)
Realism—Very Good
Complexity—9

OVERALL EVALUATION: Very Good (for hard-core enthusiasts only)

War at Sea (1976)

PUBLISHER: Avalon Hill Game Co.

SUGGESTED RETAIL PRICE: $6 (boxed)

SUBJECT: The naval war in the Atlantic during World War II.

PLAYING TIME: About one hour is needed.

SCALE: This is a strategic game using large areas rather than hexes. Turns are many months long, but units are individual ships.

SIZE: The board is only 14″ x 22″, but some additional space is needed.

BALANCE: Because they move last, the Germans have a substantial advantage. The game can be balanced by allowing the British to use inverted counters as the rules suggest.

KEY FEATURES: The central feature here is the abstract game system that has players position ships in the area they are to control. The effect is not unlike the Oriental board game *go*. Victory points arise from controlling these sea areas. A tactical flavor is achieved by having the ships in an area shoot it out one-on-one until all the ships on one side retreat or are sunk. Ships can be damaged in combat and repaired on later turns. Carrier-and land-based aircraft are handled abstractly (with little effect on the outcome), and there are special rules for U-boats and convoys.

COMMENTS: *War at Sea* is the simplest real wargame on the market and ideal for introducing newcomers to the hobby. It creates the feel of strategic naval operations and presents the player with a wide variety of choices for accomplishing his objectives. Aside from the strategic options, however, it's a die-rolling contest, and its short length makes it possible to play it often enough to get tired of it. Although it's no great shakes as a simulation, and the bane of adherents of the SPI school of realism-by-complexity, it's a pleasant little game. The full potential of the system

War at Sea, *a World War II naval game from Avalon Hill.* (PHOTO: AVALON HILL GAME CO.)

is better realized in its companion game, *Victory in the Pacific.*

EVALUATION:

Presentation—Very Good
Rules—Good
Playability—Excellent
Realism—Very Poor
Complexity—2

OVERALL EVALUATION: Good

War in the East (1974—extensively revised in 1976)

PUBLISHER: Simulations Publications, Inc.

SUGGESTED RETAIL PRICE: $26 (plastic box; also available with *War in the West* as *War in Europe* for $45)

SUBJECT: The land war on the eastern front.

PLAYING TIME: Two to seven *days* are needed for the short scenarios; one to four *months* for the full campaign. (The original version took longer.)

SCALE: This is an operational-level treatment of what would be a strategic subject. The standard unit is the division. There are thirty-three kilometers per hex, and each turn represents one week.

SIZE: The three 22″ x 33″ maps form a playing area about three-by-six feet in size. Hexes are 19mm (¾″).

BALANCE: The Soviet player has the stronger and more interesting side.

KEY FEATURES: The game is based on the *Kursk* system and features mechanized (double-impulse) movement, semirigid zones of control, overrun attacks, and unit breakdown (that is, a unit may split up into smaller units). Instead of fixed reinforcements (as in *Drang Nach Osten!*), there is an elaborate Soviet production spiral, which allows the Soviet player to produce almost any unit he wishes at a cost in production points (and, in some cases, other units). Also interesting is the use of variable Combat Results Tables: the Axis forces initially resolve combat on the most favorable CRT, but as the game goes on they must use progressively poorer CRTs. Conversely, the Soviets start at the bottom and get better.

COMMENTS: Stung by the appearance of *Drang Nach Osten!*, SPI set out to prove that they could make as unplayable a game as anyone. Or so it seemed. The original *War in the East* was an ungainly beast with two thousand counters and poorly written rules. If anyone ever finished it, the fact has gone unreported. The new version was pared down to a mere one thousand counters and redone to mate with *War in the West*. The rules were rewritten, and the result is a lot more manageable than the old one. One of the biggest differences from *Drang Nach Osten!* is the fact that the air game here is entirely abstracted. For those capable of such fine distinctions, the game is now a bit simpler and a bit more playable than *Drang Nach Osten!*, although neither is exactly suitable for a casual evening at home. The variable CRTs are a fine idea, and the production spiral gives the Soviet player a strategic flexibility. Of course, the point of a game this size is hard to imagine, although it certainly keeps people off the streets.

EVALUATION:

Presentation—Good
Rules—Very Good
Playability—Poor
Realism—Good
Complexity—8

OVERALL EVALUATION: Fair

CHAPTER 13

Yesterday, Today...and Tomorrow?

IF WE can reasonably group games that are based on hypothetical situations featuring modern military capabilities together with simulations dealing explicitly with actual battles, we can say that modern-period games were among the first commercially available wargames. In 1958 Avalon Hill's *Tactics II* pitted the military forces of Red and Blue, two imaginary nations, against each other. Each side had symmetrical armies but dissimilar terrain. With each side having nuclear capabilities, *Tactics II* was thoroughly modern; in fact, it was once used as a parody of nuclear theories of annihilation on the iconoclastic program *That Was The Week That Was* (TW3), when David Frost exhibited the game before television cameras.

In 1965 Avalon Hill produced a larger game that again matched Red against Blue. This was *Blitzkrieg*, which added a series of minor countries and named its terrain features after Avalon Hill employees (like Lake Pinsky) or popular personalities (the Great Koufax Desert). While players could adopt modern military strategies, the game introduced asymmetries in both military capabilities and terrain: Red had larger armies, while Blue's forces included more paratroops, amphibious forces, and armor. Air forces on both sides possessed nuclear weapons capability.

While popular in their day, these games had a number of difficulties, not least of which was a common absence of specific referents. Imaginary settings made the sides too hypothetical and impossible for players to identify with. They were devoid of emotional impact: fancy chess games—nothing more.

It may be worthwhile to note that the appearance of these "modern" games considerably predated the evolution of a category of reputedly historical simulations of contemporary topics. While modern wargames of all sorts can trace their lineage back to *Tactics II* (and, for that matter, to the original *Tactics*), the game's immediate successors reflected historical subjects: the events of World War II and earlier conflicts. Despite a slow start, today there are more than fifty modern-period games of several distinct varieties.

The Korean War was neither popular nor well understood by the folks at home; nor was it fought to any substantial conclusion. With the exception of its air battles, which have been incorporated as scenarios in some modern air wargames, the Korean conflict is devoid of unique interest. Although a few games have been based on the war (evidence of the extent designers will go for novelty), it's been largely a lost war for gamers as well as for the general populace.

Vietnam suffers a similar fate: politically sensitive at first, it, too, is now marked down as a lost war. Its immediacy and relevance demand attention but the subject repels players: games from *Viet Nam* to *Year of the Rat* have been uniformly unpopular and have remained in print only briefly. The style of warfare waged in Vietnam did not offer much scope for the sort of large-scale armor engagements valued so highly in modern wargames. Conventional battles were few and little known to the public. Games like *Frederick the Great,* whose object is largely to avoid combat, have never been welcomed with open arms by the majority of gamers, and military simulations of the war in Southeast Asia are no exception. Simulating the entire conflict is next to impossible: Combining unattractive ingredients like terrorism, torture, POW camps, defoliation, protective reaction strikes, mass bombings, corruption in high places, demonstrations, draft riots, and political unrest into a singularly unappetizing stew is an unrewarding business at best.

Guerilla warfare—insurgencies, terrorist operations, wars of "national liberation"—is the best known but least "gamed" form of modern conflicts. From Indochina to Kenya, from Northern Ireland to Rhodesia, from the Symbionese Liberation Army to the Black September Movement, political extremists of every stripe have attempted to use the fruits of modern technology to redirect national politics to suit their own prejudices.

This sort of thing has never been properly—or successfully—simulated. Guerilla wars are won in the "hearts and minds" of the populace; they are as much political affairs as military actions. This combination has so far defeated the best efforts of designers.

Furthermore, the choice of battlefield, target, and victim leads modern guerillas away from direct conflicts with military forces and toward the people and installations of the cities. The seats of government and the industry that supports both the economy and the government's warmaking capability are located in urban areas. To continue their lifestyle, residents of these areas are, of course, dependent on the established technology: modern communications links (like radio and telephone), modern sewage and water systems, gasoline for transportation, and—especially—electrical power.

Cars loaded with high-energy explosives, shaped-charge explosives surrounded by hardware for shrapnel, and incendiary devices in commercial packages have been used all over the world to disrupt life in major cities. The hijacking of trains and airplanes from industrial nations follows the same pattern.

Terrorist activities are not at all suited to conventional wargames. In such a simulation, the role of the Establishment player would be to continue ordinary day-to-day activities—to maintain the status quo. He would have nothing to *do,* because his military forces never get involved, and his civilian populace is generally unarmed and can do nothing but suffer and go on. The subject is as unsatisfactory in games as it is distasteful in real life. Terrorism is too real, too full of the horrors of war, too obviously not a game.

Within the realm of more conventional warfare, military capabilities in this period are of the same general type available in World War II. What has changed in the main is the tempo of war: armor, infantry, aircraft, and warships are faster, more powerful, more sophisticated. Even ground troops, better adapted to mechanized and air transport use, maneuver with a speed and a degree of flexibility unheard of in 1945. Also increased in tempo is what British General J. F. C. Fuller termed the "projectile cycle," a measure of the velocity of production and use of munitions, by which he sought to illustrate the intensity of war. Although the Vietnam War, in which several times more ordnance was expended than in all of World War II, is the grossest example of this, from the gamer's point of view the quintessential examples of modern warfare are the wars between Israel and the surrounding Arab countries.

The Six-Day War of 1967 and the October (Yom Kippur) War of 1973 were everything a gamer looks for in a war: fast-moving, decisive, purposeful—a triumph of tactics and technology. The lure of the situation and the chance to use modern armor and weaponry have proved irresistible to designers and gamers alike. Although Simulations Publications' *Sinai* (probably the only game revised because of a new war) was the first, it has had plenty of company. John Hill's *Bar-Lev* had the advantage of data from the 1973 war, but its system is unsuited to the earlier conflicts. That the most recent entry in the Suez sweepstakes, *The Arab-Israeli Wars*, is a third-generation *PanzerBlitz* says much about the attractions of the system, the quality of the game, and the way gamers see the war.

The other major grouping of games in this chapter concerns the possible conflicts that might arise in the next decade or so. The next "war to end war" might actually achieve that end—by eliminating the warriors, along with most life above the level of cockroach and lichen. A full-scale nuclear exchange between two or more superpowers would be, at best, no more than Lloyd George's proverbial "blood-stained stagger to victory."

While there are games that consider such a prospect, the average gamer prefers his horror in rather restricted doses. Nuclear weapons options are present in many games of the contemporary era. Even the most serious simulations of modern warfare, many of which contain at least tactical nuclear devices, apply them relatively bloodlessly. The destructive power of nuclear weapons is restricted to the battlefield; targeted units are destroyed, but little or no consideration is given to side effects beyond the immediate irradiation of the area. The psychological impact on front-line troops is given even less consideration than the medical effects. The thrust of nuclear games and systems is away from the more horrible reality—which probably cannot be modeled adequately without incendiary maps and counters.

Nonetheless, gamers are curious about the possible results of contemporary conflicts, from the Middle East (*Oil War*) to the Far East (*The East Is Red*), on land (*Mech War '77*), in the air (*Air War*), and on the seas (*Sixth Fleet*). So far, this area is almost the exclusive province of Jim Dunnigan and his staff at SPI.

A common difficulty shared by all these games is the lack of hard data. The common assumption, for instance, is that better tactics and the far more sophisticated technology of NATO are at least a match for the numerical superiority of the Warsaw Pact forces in a land battle in modern Europe—but there's no way to know for sure. This has not diminished the popularity of the large number of games dealing with such a conflict.

Air games typically feature many scenarios with a variety of aircraft on various missions. Since there has been combat between modern aircraft, there is somewhat more data available, but *Foxbat & Phantom*, the first of these games, was not particularly accurate. It was, however, playable, which is more than can be said for its successor, *Air War*, which carries realism to absurd extremes—sixty pages of rules and ninety pages of tables for plane-to-plane combat!

Naval games seem to be based as much on guesses as on research. This is understandable, perhaps, in view of the classified nature of much of the relevant material. *Sixth Fleet* has been described as "*Napoleon at Waterloo* goes to sea," but all of these games have been consistently—if not uniformly—unrealistic.

In all settings, modern warfare—with its nuclear devices, laser guidance systems, electronic countermeasures (ECM), surface-to-air missiles (SAMs), wire-guided rockets, detection gear, helicopter gunships, spy satellites, computers, and so on—is far more complicated than ever before, and simulations of it reflect that level of complication. Few are suited for any but the experienced, sophisticated gamer. For some, however, the range and variety of weaponry offers a unique challenge.

While the uncertainties of modern warfare preclude definitive answers, it's more fun to speculate about the outcomes of hypothetical conflicts than to find out the hard way.

Evaluations

Air War (1977)
PUBLISHER: Simulations Publications, Inc.
SUGGESTED RETAIL PRICE: $15 (plastic box)
SUBJECT: Modern tactical air combat.
PLAYING TIME: The game takes one to two hours.
SCALE: This is a tactical simulation; units are individual jet aircraft. Hexes are five hundred feet across (with 250-foot altitude levels), and turns are equal to two and one-half seconds.
SIZE: A large table is needed. There are eight 8½" x 11" geomorphic maps, and additional space is required for Aircraft Control Displays. Hexes are 16mm (⅝").
BALANCE: Excellent.
KEY FEATURES: The game is a much more complicated successor to *Foxbat & Phantom*. The outstanding feature (other than the size of the rule book) is the turn sequence. Movement is sequential, not simultaneous, but aircraft with advantages of facing, altitude, or trailing position get an opportunity to preserve their superior position by moving *last*. The operations of movement are involved and heavily dependent on the Aircraft Control Display for workability. There are separate markers and/or tracks for speed, throttle setting, climb/dive and roll attitude, acceleration, and turn progress—among others. There are complex considerations

for speed change due to climb, dive, or roll; the effect of banking on turns; and the relation of speed changes and attitude. Aircraft on the map may be faced at hexsides or the junctures of two sides. Combat is involved but not to the degree of movement. All sorts of different cannon and air-to-air missiles have been put in the game, all judged for effectiveness in more than one way. A ground attack system is included, but it's *oversimplified* (incredibly enough) and unrealistic.

COMMENTS: There is apparently no absurd extreme to which SPI will not go in its quest for realism (or the appearance thereof) for which it will not have a small but vocal coterie shouting huzzahs at the results. Witness this game and its ludicrously high ratings in *Strategy & Tactics* magazine. If you took a semester course in the rules, you would probably find this a rewarding and even somewhat playable simulation. You would also find that many of the systems are unnecessarily complex and unwieldy, even for the intended level of realism. Since complications enhance the *image* of accuracy, perhaps no one thought it worthwhile to develop the thing adequately. Another annoyance is that some tremendously complex systems are based on what is, at bottom, nothing more than educated guesswork; the data simply can't be verified. There are some nice features here, but most people will do better to await their adoption by less ambitious efforts.

EVALUATION:
- Presentation—Good
- Rules—Good
- Playability—Fair to Poor
- Realism—Excellent (maybe)
- Complexity—10

OVERALL EVALUATION: For fanatics only

The Arab-Israeli Wars (1977)

PUBLISHER: Avalon Hill Game Co.

SUGGESTED RETAIL PRICE: $12 (boxed)

SUBJECT: Land combat in the Arab-Israeli conflicts of 1956, 1967, and 1973.

PLAYING TIME: About three to four hours per scenario.

SCALE: This is a tactical-level game using companies and platoons. Each game represents about six minutes. Each hex represents 250 meters.

SIZE: There are four 8″ x 17″ geomorphic boards that may be used singly or in various combinations, depending on the scenario. Hexes are 19mm (¾″). Most scenarios can be played on any normal-sized table.

BALANCE: This varies according to the scenario, but is generally good.

KEY FEATURES: The standard modern-era tactical system introduced by *PanzerBlitz* is in evidence here. The Combat Results Table is odds/ratio, with elimination of units as well as dispersals. Movement is sequential with some interdiction. Each unit is rated for a variety of combat capabilities. Morale rules are optional, as are many of the special unit rules covering helicopters, bridge-building, and so on.

COMMENTS: This is the second descendant (following *Panzer Leader*) of the famous *PanzerBlitz*, and while it is not in some ways as successful or as satisfying as its preeminent forefather, *The Arab-Israeli Wars* is nonetheless a good tactical game. Many of the earlier rules have been updated and expanded, and play in this game flows much smoother than in *PanzerBlitz*. The main problem is that the system is becoming a little tired, and players may feel they are just getting some new scenarios for an old game. While some of these scenarios are exciting, many are dull and unwieldy. Moreover, according to many experts in this area, a surprising amount of the hard information is of doubtful veracity. SPI's *October War* is more accurate and innovative but is, unfortunately, less successful overall.

EVALUATION:
Presentation—Very Good
Rules—Good
Playability—Good
Realism—Good
Complexity—7

OVERALL EVALUATION: Good

Bar-Lev (1974)

PUBLISHER: Conflict Game Co.

SUGGESTED RETAIL PRICE: $12.75 (boxed)

SUBJECT: Land and air combat in the Yom Kippur War of October, 1973.

PLAYING TIME: Eight to ten hours are required.

SCALE: This is, in effect, a combination of strategic and tactical games. Also, the scale on the Syrian front is half that of the Suez (Egyptian) front. Each turn represents one day. Israeli units are battalions; Arab units are regiments.

SIZE: The 33″ x 22″ mapsheet uses large, 21mm (⅝″) hexes. Additional space is required for charts and tables.

BALANCE: With the air module, Israel has an edge; without it, balance shifts toward the Arab player.

KEY FEATURES: *Bar-Lev* uses a split board to simulate both the Suez and Syrian fronts; the Israeli player (only) can shuttle units back and forth between fronts by means of transit boxes. Each day of combat includes no less than twenty different phases. There is a vast array of unit types and provisions for ranged artillery, aircraft, helicopters, Arab SAMs, bridge-building, fortifications, morale, and much more. The air rules are modularized and may be omitted from the game to reduce complexity—but that also affects balance. Victory on a given front is based on territorial objectives; to win the game, you must win on one of the two fronts and at least draw on the other.

COMMENTS: This is an ambitious game—perhaps too much so. The tactical system would work well for *one* of the fronts. The strategic options and air rules would significantly enhance a game in which the combat on both fronts was more abstracted. Combined as they are, they make a package that is forbiddingly complex for all but quite experienced gamers. Nonetheless, despite the rules, which are (in the original edition, at least) casually chatty and hopelessly ungrammatical, many people swear by the game. If you are a wargame veteran interested in the subject matter, and if you spend a few days soaking up the rules and the routine of play, you'll be rewarded with an enjoyable, challenging, and—except for its length—reasonably playable game. If you're not willing to go to that length, *Bar-Lev* is just an incredible stew of confusing detail better left unconsumed.

EVALUATION:
Presentation—Fair
Rules—Poor to Fair
Playability—Fair
Realism—Good

Complexity—8

OVERALL EVALUATION: For hard-core fanatics, Very Good; for normal people, Fair

FireFight (1976)

PUBLISHER: Simulations Publications, Inc.

SUGGESTED RETAIL PRICE: $14 (plastic box)

SUBJECT: Contemporary land combat in Europe.

PLAYING TIME: One to two hours for the introductory scenarios; much longer for the more complex ones.

SCALE: This is a close-tactical game in which counters represent two to five men or one vehicle. Hexes are fifty meters across, and turns are forty seconds long.

SIZE: The two 22″ x 34″ maps use 16mm (5/8″) hexes. A bit of additional space may be required, but then some scenarios require only one of the maps.

BALANCE: The American/NATO forces are greatly favored.

KEY FEATURES: Within the three levels of complexity, everything is included: line-of-sight rules, fire during movement, indirect fire from off-board artillery, dummy counters, smoke shells, mines, overwatch and suppressive fire, special AFV rules, improved positions, overruns, and even provisions for future (hypothetical) weapons system developments. The scenarios become progressively more complex and are designed to lead the player into the game systems gradually. Victory conditions are set up so that the Soviet Player must behave in accordance with official Soviet tactical doctrine. A twenty-page reference booklet is also included.

COMMENTS: *FireFight* is an extremely professional, accurate, and well-laid-out piece of work. It is, however, a learning device—not a game. *FireFight* was specifically designed for the U.S. Army to use for training purposes; the aim of the system was realism, not playability. The scenarios that can be played without an enormous investment of time and effort are rather wooden and one-dimensional, and the constraints on the Soviet player make playing that side less than satisfying. Using all the rules and the most complicated scenarios makes a more interesting but also far longer and more complex game. *FireFight* is possibly the most tedious game on the market; every single unit must check for defensive opportunity fire every single hex on every single turn. If enjoyment is a consideration in your game playing, pass this by.

EVALUATION:

Presentation—Excellent

Rules—Very Good

Playability—Poor

Realism—Excellent

Complexity—8

OVERALL EVALUATION: As a simulation, Excellent; as a game, Poor

Foxbat & Phantom (1973)

PUBLISHER: Simulations Publications, Inc.

SUGGESTED RETAIL PRICE: $14 (boxed; less in other packaging)

SUBJECT: Modern air combat.

PLAYING TIME: Thirty minutes to two hours.

SCALE: This is a tactical simulation; units are individual aircraft. Thousand-meter hexes and thousand-meter altitude levels are used. Each turn represents thirty seconds.

SIZE: The 22″ x 34″ map uses 16mm (5/8″) hexes.

BALANCE: This is good.

KEY FEATURES: The design draws upon techniques developed in air games of earlier periods—for example, SPI's now-defunct *Flying Circus*—to re-create the elements of jet-fighter combat. As a simplification, the capabilities of aircraft (which, of course, vary greatly) have been designed around their performance at altitudes above seven kilometers. The turn mode simulates the turning radius of a jet by requiring it to move a certain distance in a straight line before turning one hexside. The climb rate is dependent on speed (like the turn mode) and an airplane's ability to accelerate. Electronics are only abstractly present, and the methods of combat—cannon, heat-seeking or radar-homing missiles—are also fairly simple and abstract.

COMMENTS: Of those produced when SPI went on a tactical air games spree in the early 1970s, *Foxbat & Phantom* is the most successful—and the sole survivor. In an admittedly limited way, the game vividly depicts some of the elements of modern air combat. The MiG-21 Foxbat is a "horizontal" fighter that is able to turn tightly even at high speed. The F-104 Phantom is a "vertical" fighter that is able to climb and dive rapidly and utilize the vertical dimension for maneuver. Unfortunately, the height restrictions of the design and some inaccurate assessments distort the capabilities of various aircraft. Electronics has been underplayed for simplicity, and the turn mode is no more than a playable abstraction. The turning radius of a modern aircraft is limited more by pilot stamina than by mechanical factors, and in a critical situation a pilot will push to his limits rather than take a shot from an opponent. Nonetheless, the game is fun and far more playable than its "successor," *Air War*.

EVALUATION:
Presentation—Good
Rules—Good
Playability—Excellent
Realism—Fair to Good
Complexity—5

OVERALL EVALUATION: Good

Mech War '77 (1975)

PUBLISHER: Simulations Publications, Inc.

SUGGESTED RETAIL PRICE: $14 (boxed)

SUBJECT: Hypothetical land engagements in Germany, the Middle East, and China, utilizing current weapons systems or those under development.

PLAYING TIME: The game takes six to eight hours.

SCALE: This is a tactical simulation; units represent platoons, companies, or batteries. Each hex equals two hundred meters, and a turn represents one to six minutes.

SIZE: Two 17″ x 22″ mapboards butt to form a 22″ x 34″ playing surface. Hexes are 16mm (5/8″).

BALANCE: Although balance varies from scenario to scenario, the advantage generally rests with the defender.

KEY FEATURES: The use of sequential movement and simultaneous combat (a system shared with its sibling, *Panzer '44*) is a compromise between playability and realism. Written firing plots and the weak command control system are also shared with *Panzer '44*. There is indirect fire from off-board artillery, and overruns are provided for, though these are risky at best. Helicopters are used for transport, spotting, and direct attacks.

COMMENTS: *Mech War '77* demonstrates the

lethality of modern weaponry in dramatic fashion. Infantry and even some armored personnel carriers can blow away armored units with deadly consistency (a distinct improvement over the weak infantry of *Panzer '44*). Tactical finesse becomes important, as direct actions tend to result in great slaughter for the impetuous. On the negative side are the same written firing plots and the same stupid command control rules of the World War II relative. *Mech War '77* was a replacement for the old *Red Star/White Star* (not to be confused with the *new* version) and will probably be phased out in favor of its successor, *Mech War 2*. (If at first you don't succeed . . .) But it can be fun for armor buffs.

EVALUATION:
- Presentation—Good
- Rules—Good
- Playability—Fair to Good
- Realism—Very Good
- Complexity—8

OVERALL EVALUATION: Good

MiG Killers (1977)

PUBLISHER: Gamescience

SUGGESTED RETAIL PRICE: $8 (resealable plastic bag)

SUBJECT: Jet combat from 1945 through the 1980s.

PLAYING TIME: Thirty minutes to two hours.

SCALE: This is a tactical simulation using individual aircraft counters. Hexes and altitude levels represent a distance of 245 meters in each dimension; turns equal five seconds.

SIZE: The 23" x 34" map uses 16mm (⅝") hexes.

BALANCE: This is fine.

KEY FEATURES: The somewhat artificial turn mode has here been discarded in favor of a method related to the g-forces that a pilot would experience in making a turn. Turn angles are chosen freely and limited only by the physical stress capacity of the human body or the aircraft. Rather than the displays with markers used by its better-known competitors, the climb/dive and turn rules require a fair amount of bookkeeping. Instead of indulging in *Air War*'s speculation, the game has dealt with the relative effectiveness of air-to-air weapons fairly abstractly. There are provisions for ECM, fuel, ace status, and so on. More than forty scenarios are included.

COMMENTS: *MiG Killers* may be the best tactical jet-combat game going. More than any of the others, its clever mechanics capture the essence of plane-to-plane combat in the modern era. The turn rules are a major advance in realism and eliminate the "cheap shots" that the turn mode allows pursuing pilots. Unfortunately, the game is little known, and the competitors, especially *Air War*, have far more going for them in packaging and promotion. It, however, does have its drawbacks. Some gamers will object to the bookkeeping required, and realism nuts will balk at some of the abstractions in the combat rules, though they make the game far more playable than a monster like *Air War*. The rules could stand proofreading—if not rewriting—and some of the explanations lack lucidity. Furthermore, the physical standards of the game are below those of the major companies. Newcomers will have trouble deciphering the rules, but veterans should find this an attractive alternative to the other air games on the market.

EVALUATION:
Presentation—Fair to Poor
Rules—Fair
Playability—Very Good
Realism—Very Good
Complexity—5
OVERALL EVALUATION: Good

Oil War (1975)

PUBLISHER: Simulations Publications, Inc.

SUGGESTED RETAIL PRICE: $10 (boxed); $4 (plastic envelope)

SUBJECT: Three hypothetical battles fought in the 1970s in the Persian Gulf area, involving European, American, and Arab land and air forces.

PLAYING TIME: The game takes two to three hours.

SCALE: This is an operational-level simulation. Land units are brigades, and air units are squadrons. Each turn represents days; each hex thirty kilometers.

SIZE: The 17″ x 22″ mapsheet uses 16mm (5/8″) hexes.

BALANCE: American air superiority is balanced by the victory conditions.

KEY FEATURES: *Oil War* uses one of the more novel and innovative game systems around. Ground movement is fast and uniform for all units, and there are a lot of "drop-in" units: paratroopers and others transported by air. Air transport points measure air capacity each turn for transport and supply. The game centers on control of airfields (needed as bases for aircraft) and oil wells (for victory points). A lot of air units that can move just about anywhere on the map contribute most of the offensive punch, either in air combat or as close support for friendly ground attacks. Although attacking is voluntary, ground units must stop when entering enemy zones of control but can leave the following turn. A relatively bloodless differential Combat Results Table forces a player to surround units he wishes to destroy. Three different scenarios are provided.

COMMENTS: *Oil War* has never received much notice in wargaming circles. Perhaps the situation portrayed is too discomforting to be popular. The game system, however, is clean and fast, and the situation—a nice balance of American air superiority and victory conditions that emphasize the importance of the Arab land superiority—is fluid and exciting. Although there are a lot of things going on in the game, it is still simple to learn and play. On the other hand, the portrayal of naval air units is unrealistic in regard to both the particular aircraft capabilities and the numbers shown—unless we are to assume that the entire U.S. Pacific carrier force is cruising in the Gulf of Oman!

EVALUATION:
Presentation—Good to Very Good
Rules—Very Good
Playability—Very Good
Realism—Fair
Complexity—5
OVERALL EVALUATION: Good

Revolt in the East (1976)

PUBLISHER: Simulations Publications, Inc.

SUGGESTED RETAIL PRICE: $4 (plastic envelope)

SUBJECT: A hypothetical revolt by Warsaw Pact nations in the immediate future. Additional scenarios present variations of the Hungarian revolt of 1956 and the Soviet invasion of Czechoslovakia in 1968.

PLAYING TIME: The game takes two to three hours.

SCALE: This is a strategic-level game using corps and armies. Each hex equals fifty-six kilometers, and each turn is a week.

SIZE: The 17″ x 22″ game map uses 16mm (5/8″) hexes.

BALANCE: The Soviet units are stronger, but the NATO player has a slight edge.

KEY FEATURES: Although it conveys the illusion of a very complex situation, the mechanics of *Revolt in the East* are fairly simple. The standard Move/Attack sequence is used, but there are some very interesting variations in the types of movement available. Ground movement costs vary not only by terrain but by country. Air units aid conventional attacks or may engage in a simplistic form of air-to-air combat. The greatest novelty of this game is the Warsaw Pact Revolt Table, which determines the revolting countries and NATO intervention (which allows direct military action by the NATO player) and provides the Soviet player with his greatest challenge.

COMMENTS: The folio format produces short playing times and uncomplicated rules, a boon to casual and beginning gamers. The Soviet player must crush revolts immediately or face unpleasant consequences from the Revolt Table. The NATO player, on the other hand, must content himself with moving the small national armies of the revolting nations until the Soviet position degenerates enough to trigger direct NATO intervention. While it makes the game too dependent on luck for some tastes, the randomness built into the game in the form of the Revolt Table adds variety to the game and preserves its novelty over repeated playings. A lively and pleasant diversion.

EVALUATION:

Presentation—Good
Rules—Very Good
Playability—Good to Very Good
Realism—Fair
Complexity—5

OVERALL EVALUATION: Good

Sixth Fleet (1975)

PUBLISHER: Simulations Publications, Inc.

SUGGESTED RETAIL PRICE: $10 (plastic box)

SUBJECT: Modern naval (and air) warfare in the Mediterranean Sea.

PLAYING TIME: The game takes three to six hours.

SCALE: This is an operational-level simulation. Units are groups of twelve planes, three destroyers or conventional submarines, or larger individual ships, including nuclear submarines. Each hex represents fifty miles; each turn represents eight hours.

SIZE: The 21″ x 32″ game map uses 19mm (3/4″) hexes.

BALANCE: Air superiority gives NATO forces a bit of an edge.

KEY FEATURES: Despite a certain complexity—largely a matter of the number of choices that a player must make from a *very* large number of options—*Sixth Fleet* is relatively simple to play. The unusual turn sequence has combat preceding movement. Units have different attack strengths against submarines (which can retreat before combat), surface ships, and aircraft. Combat is dependent on ECM, and the Combat Results Table is relatively "bloodless."

COMMENTS: This is an enjoyable and interesting game that is nonetheless quite peculiar in its assumptions. The tactics required would be more at home in a Napoleonic game than in a modern naval simulation. The outcome is very dependent on the setup; one misplaced unit—one hole—can lead to the loss of an entire air force. Not only does this slow play considerably, but it also forces the players to take extreme care about something that has no basis in reality. Furthermore, the weapons systems estimates are quite unrealistic. Although its relative anonymity is probably deserved, *Sixth Fleet* misses being quite a good game by a small amount on a lot of points.

EVALUATION:
Presentation—Good to Very Good
Rules—Good
Playability—Good
Realism—Fair
Complexity—6

OVERALL EVALUATION: Fair to Good

Strike Force One (1975)

PUBLISHER: Simulations Publications, Inc.

SUGGESTED RETAIL PRICE: Free (envelope)

SUBJECT: Contemporary land combat between U.S. and Soviet forces in West Germany.

PLAYING TIME: Fifteen to twenty minutes.

SCALE: Call it an operational game; the units are companies (it says), and the scale is anyone's guess.

SIZE: The map and Combat Results Table sheet is 6″ x 10½″ and uses 16mm (⅝″) hexes.

BALANCE: This is almost entirely dependent on the luck of the die (provided).

KEY FEATURES: This is your *basic* wargame. Units have no movement or combat factors *per se,* and since they are identical, the simple Eliminate/Retreat CRT can be considered either odds/ratio or differential. Zones of control are what SPI, at least, would call "Rigid" and "Inactive." Combat is optional; movement out of zones of control is allowed but prohibited within them. Aside from the three town hexes, the occupation of which is the object of the Soviet player, the only terrain feature is five impassable woods hexes. Optional rules allow stacking and movement through woods at a penalty.

COMMENTS: If you were wondering what a Complexity 1 game is, this is it! It's specifically designed by SPI as an introductory wargame, but it's flawed as an introduction, and it can hardly be considered a wargame. It's too simple to be anything but a very short die-rolling contest; there's no opportunity for skillful play, and with a total of ten units and only four turns, there's not enough time for lucky die rolls to even out. No one—regardless of intelligence or experience—would play this more than once. The zones-of-control rules are wrong for the complexity level; allowing movement out of but not through zones of control is just the sort of fine distinction first-timers have trouble grasping. The original, basic *Napoleon at Waterloo* was a good introductory game that was playable (and replayable), but with the growing popularity of the modern period, SPI evidently decided to replace it with this dud. But it's free—and worth every penny.

EVALUATION:
Presentation—Very Good
Rules—Very Good
Playability—Excellent
Realism—Nonexistent

Complexity—1
OVERALL EVALUATION: For utter novices, marginal; for anyone else, forget it

Tactics II (1958)
PUBLISHER: Avalon Hill Game Co.
SUGGESTED RETAIL PRICE: $6 (boxed)
SUBJECT: Land combat between two imaginary countries equipped with contemporary forces.
PLAYING TIME: The game takes two to five hours.
SCALE: This could be termed a strategic or operational game. The time/distance scales are unspecified, but the basic unit is the division.
SIZE: The 22″ x 28″ game board uses half-inch *squares*.
BALANCE: *Too* good; it's almost impossible to win or lose.
KEY FEATURES: This (along with its original version, *Tactics*) is where it all began. The odds/ratio Retreat/Eliminate/Exchange Combat Results Table was the standard for subsequent games, as was the simple Move/Fight turn sequence. The obvious unique feature is the use of squares instead of the hexes that replaced them. There are provisions for road and hidden movement and special rules for Amphibious, Mountain, and Paratroop units. Headquarters units have no function whatsoever unless the nuclear option is used. Other options include replacements, isolation from supply, and weather effects.
COMMENTS: Aside from its historical importance, this game has no redeeming qualities. The imaginary countries of Blue and Red fail to engage our sympathies, and the situation is not just unattractive; it's hopeless. Against an even vaguely competent opponent, it can't be won. The mobility of the road network and the mountain and woods bottlenecks combine to produce an inevitable stalemate directly across the center of the board, leaving a massive—and suicidal—frontal assault as the only way out of the deadlock. It is justified as an introductory game, but the number of special rules, the absence of hexes, and its inability to be played to a satisfactory conclusion give the lie to this pretext. *Tactics II* is overdue for retirement
EVALUATION:
Presentation—Fair to Good
Rules—Fair
Playability—Good
Realism—Poor
Complexity—4
OVERALL EVALUATION: Poor

Yalu (1977)
PUBLISHER: Conflict Game Co.
SUGGESTED RETAIL PRICE: $11.98 (boxed)
SUBJECT: The Chinese counteroffensive in Korea from November, 1950, to May, 1951. The conflict is mostly on land, though there is an air element.
PLAYING TIME: The game takes two to five hours.
SCALE: This is an operational game. Units are battalions, regiments, and divisions. Hexes are ten miles across, and turns represent the passage of one week.
SIZE: The 19″ x 25″ game board uses 19mm (¾″) hexes.
BALANCE: Fairly good (experience favors United Nations forces; skill favors the Chinese).
KEY FEATURES: The outstanding element of the game is the way two completely dif-

ferent game systems—for supply, movement, and combat—have been merged to give the United Nations and Chinese forces the peculiar capabilities each had. The more flexible Chinese are given great mobility across rough terrain and through UN zones of control but are restricted by limited supply. The UN forces have the advantages of air and naval bombardment and much better supply, but they are limited in mobility and can be crippled by weather. The Chinese have more units and can move them more freely, while the numerically inferior UN units (coupled with the bombardment capabilities) have more punch.

COMMENTS: Designer John Hill is dedicated to the proposition that the *feeling* of a situation is the quantity a designer must seek the most in a game. The common criticism, therefore, is that reality is sacrificed to form—not always with any good result. *Yalu* distorts geography (in the mapboard) and history (in the appearance and strengths of various units). Some cavils have been raised about minor provisions of the rules as they affect unit abilities, too. Despite its inaccuracies, it *feels* realistic. *Yalu* is reasonably balanced, fast-moving, and fun.

EVALUATION:

Presentation—Fair
Rules—Good
Playability—Good
Realism—Good
Complexity—7

OVERALL EVALUATION: Good

CHAPTER 14

Science-Fiction Games: The Stars and Other Destinations

WHILE HISTORIAN gamers may be content to look backward, most others, reasonably enough, are curious about where they are going. The possible conflicts of the future are the province of science fiction.

Back in the early days of science fiction, war in the future meant war in space—and on a grand scale. In the stories of E. E. "Doc" Smith and Edmond "World-Wrecker" Hamilton, thousands of mile-long spaceships hurled "unimaginable" (but, of course, imagined) torrents of energy rays at the protective force fields of equally vast fleets of superdreadnoughts, which replied in the same manner. Nothing material, of course, could withstand such an onslaught, and even the energy screens could not prevent scores of ships from being annihilated in microseconds. Moons, planets—even suns—were snuffed out like wet matches.

In the most common variation, one superpowered spaceship, designed and "flown" by a handful of brilliant scientist-inventors (*always* from Earth), challenging the might of nations or worlds, would battle flying, armor-piercing monsters and aerial fleets in the skies of alien planets. At first, steel armor (a mere four feet thick!) and machine guns (perhaps "souped up" with nonradioactive tactical nuclear bullets) were standard equipment, but that lasted only through the initial skirmishing. With the manufacturing capacity of friendly natives, a bit of insight into local custom, and a modest application of thought, a ship like the *Skylark* would get overhauled at the flip of a page. New devices were devised in seconds, jerry-rigged in minutes, and manufactured in hours (with apologies from the local help about the unavoidable delays). Heat rays, X-rays, blinding light, subsonic sound, Z-rays, Q-rays, tractor-pressor beams, zones of impenetrable force, total conversion of matter into energy—a secret weapon was unheard of one day, invincible the next, and obsolete a week later. It was tough to compete in those days!

In an era of fission and fusion bombs and an amazing variety of lasers, tales of such death rays and mass-conversion explosives seem less outlandish than they did when they were written (a decade or so before Hiroshima, in many cases) or during the smug 1950s, when "everyone knew" heat rays were impossible and an atomic bomb was—and always would be—*the* ultimate weapon. On the other hand, questions of economics and logistics were ignored or treated with childish superficiality; the impracticality of fleets of expensive, gargantuan—but effectively fragile—battleships universally escaped notice. Strategy equaled force, and tactics meant a new secret weapon. Battles were as militarily realistic as Hopalong Cassidy's infinite repeater.

The period of 1939–1955 (the lifespan of the archetypal pulp magazine, *Planet Stories*) saw the introduction of better writing, more sophisticated plots, and a somewhat more varied and believable cast of characters. In most of the pulps, however, it was the glorious age of space opera. Fleets might thunder in the background, but the interest was in the fate of the princess and her champion. Heroes were almost always heroic (and usually nonscientists); damsels were beautiful (and always in distress); and villains were greedy and despotic rather than just evil. Anachronistically—since movies and television tend to run about twenty-five years behind the literature of science fiction—the best-known examples of this sort of thing are the wonderful *Star Wars* and the conspicuously bad *Battlestar Galactica*. Not too plausible, but the sort of glorious adventure that has been entertaining the human race for three thousand years.

The science fiction of the last quarter-century has stayed closer to more logical extensions of contemporary combat and has thus been far more believable. The development of the atomic bomb, the jet airplane, and the ICBM did not eliminate ships, tanks, hand weapons, and individual soldiers, and there is no reason to presume that the invention of the starship will do so. In short, future combat in contemporary science fiction tends to be on a scale and of a scope we can grasp; it involves people with whom we can identify; it is fought in ways that make sense to us. Science-fiction games reflect these more modern trends.

Robert A. Heinlein, Poul Anderson, Keith Laumer, H. Beam Piper, Andre Norton, Jerry Pournelle, David Drake, Gordon R. Dickson, and Joe Haldeman, among others, have all dealt plausibly with war in the future. Although many emphasize the combative nature of man, the presence of war in science fiction—and in science-fiction games—is as much a practical matter as a philosophical one. Conflict is essential to all fiction. Overt conflict is the easiest and most obvious sort for the author to handle and for the reader to grasp. And the variety of weapons, battle craft, terrain, and types of combat allowed by alternate extrapolations into the future is a bonanza for fiction and gaming alike.

Of course, many subjects and themes of science fiction are not suited to games at all, and even more do not fit a wargame format. Others simply haven't been tried or have been given short shrift. The paradoxes and complexities of time travel that make it so fascinating but tricky to handle in fiction have thus far proven too great for wargames. Its treatment in some "fringe" games has been severely limited, but some approaches seem possible, particularly if computers are used.

Alternate universes—parallel time tracks—have been used little. The specific settings chosen may lack sufficient appeal and the instant recognition that war in space provides.

Mental powers—psionics—are periodically in fashion in science fiction but have been ignored in gaming except for *StarForce*, a game that sticks to fairly gross uses: moving an entire starship, for example.

Handling precognition or telepathy—to name just two standard examples—would clearly be extraordinarily difficult.

Harder to understand is the absence of games dealing specifically with the interface between ground and space tactical games—simulations of the invasions or raids of H. Beam Piper's *Space Viking*, for instance. The specifically military nature of a planetary invasion, the variety of units (tanks, planes, and starships in a single battle!), and the complexity of factors involved all seem to cry out for a wargame.

Beginning with Robert Heinlein's *Starship Troopers* in 1959 and Gordon Dickson's *Dorsai!* the following year, man-to-man tactical combat has achieved a secure place in contemporary science fiction. Indeed, it is now more typical of military science fiction than the old standby, space (ship-to-ship) combat.

In Dickson's future, electronic countermeasures (ECM) have developed to such a point that fighting has once again become fairly primitive: the typical infantry weapon fires a tiny needle by mechanical means—a sort of deadly BB gun. This seems to be too dull for gamers, who have taken instead to the pattern of Heinlein's Mobile Infantry; troopers have power-assisted, jet-equipped armor (half spacesuit, half medieval plate), and are armed with hand flamers, grenades, missiles, and miniaturized nuclear weapons. Each "infantryman" combines aspects of all the traditional army branches: artillery, cavalry, armor, and foot soldier. This depiction, echoed in Joe Haldeman's novel *The Forever War*, is the focus of several games, such as *Starship Troopers* and *StarSoldier*, and a feature of nearly everything whose subject and scale are appropriate.

Quite a few games focus on tactical clashes on "planetary" surfaces: Antarctica, Mars, asteroids, the moons of another planet, the worlds of another sun. These resemble tactical games of more conventional subjects; their attraction is mostly of detail and flavor. The tanks may be robotic, enormous, and powerful; the weapons or modes of transportation exotic; the terrain different in quality, appearance, or shape (the map of *Black Hole* simulates a torus—a doughnut); everything is a step or two beyond what we have now.

The fact that the rules and situations can be fairly conventional has disturbed some people, but such objections are akin to statements that science fiction should be written in some bizarre or unusual style or language that would be more "appropriate" to its futuristic subjects. The practical answer to both sorts of criticism is the same: the reader or gamer can only assimilate a certain amount of difference or change at a time. Treating unusual subjects and themes in wildly unconventional ways typically results in incomprehension and frustration, and a consequent resounding rejection in the marketplace.

Spacecraft have all the attractions of tanks —in spades. The epitome of functional design, they evoke not the carnage and horror of war but the glorious thunder of battle. As treated by authors like Poul Anderson and H. Beam Piper, spaceship combat is clearly descended from the naval battles of the past, and the relatively few games devoted to tactical (ship-to-ship) space combat are as obviously related to *Jutland* and *Wooden Ships & Iron Men*. Thanks to *Star Wars*' pyrotechnics and a popular enthusiasm for lasers, ray guns are back in fashion. Standard starship armament, in fiction and games, typically includes such beam weapons, missiles, and some sort of armor ("collapsium," Smith's marvelous "arenak," or old-fashioned steel), defensive "cloaking" device, or shielding force, which either reduces potential damage or lessens the probability of being hit by enemy fire.

The strong influence of *Star Trek*, the television series as well as the multitude of computer games derived from it, shows up in the mechanics of play: rather than having fixed, independent movement and combat factors, a unit is limited to a certain amount of energy per turn, which activates movement and weapons systems alike, generally up to some specified maximum. This same sort of thing, in somewhat different guise, shows up in *StarSoldier*.

Following current fictional practice, most tactical games emphasize tanks, GEVs, armored infantry, and the like rather than starships, which are treated most commonly as the operational arm in what are basically strategic games whose objects tend to be not just the destruction of enemy ships but the control of certain jump routes (*Imperium*), warp lines (*WarpWar*), or stargates (*StarForce*), and the occupation of worlds and bases. These fall into two broad groups: two-player wargames and multiplayer games of interstellar expansion and conquest.

The subject of the first group is quite literally a war: not a battle and not a series of conflicts spanning generations. Almost always this is a war for political hegemony (a popular cause of war in and out of fiction), and very often it involves some variation of the underdog "good guys" versus a repressive government of bad guys. Simulations Publications' new game, *Freedom in the Galaxy*, seems to be based rather distinctly on the plot background of *Star Wars*, now the most well known example of this sort of thing but hardly the only one. *BattleFleet Mars*, which is set entirely within the solar system, echoes most strongly the situation of Heinlein's *Red Planet*, but also, to a lesser degree, dozens of other instances in which colonists fight for independence from the mother planet. *Imperium* pits a young, expansionist Terran confederation against the provincial governor of a vast, established, and somewhat indifferent galactic empire.

Despite numerous practical objections, galactic empires (or, more accurately, interstellar political groupings) remain popular in fiction and games. There are sound reasons for this, some thematic. A nice, autocratic empire makes a perfect foil for

Cover of Metagaming's science-fiction and fantasy-game magazine The Space Gamer.
(PHOTO: METAGAMING CONCEPTS)

struggles for independence and, not so incidentally, for a good deal of philosophy about the value of freedom and the desirable nature of government. As the big government to end all big governments, a galactic empire gloriously exemplifies everything that everyone dislikes about government. The fact that a lesser example doesn't make the point as effectively or at all is precisely the answer to other objections.

The formation of galactic empires—the grand days of exploration, discovery, colonization, exploitation of resources, alliances, betrayals, and wars—is the subject of a final set of games. These are vast in scope: *Outreach* gets fifteen hundred light-years to a hex and entire generations into one turn! Because of the desirability of limited intelligence in terms of planetary locations and resources and player identification, location, and intent, these games are perhaps best played with the use of a large computer, but there are certainly plenty of intriguing board game versions.

This has always been a popular form but it has a few difficulties. Although most such games accommodate a varying number of people, they ought to be played with at least three and preferably four or more. As two-player games they have ample strategic scope, but they lack two of their most attractive elements: surprise and the interaction between players (the very basis of a game like *Diplomacy*). Raising the level of a player's civilization may be an end in itself or simply a means of pursuing another aim (most commonly the accumulation of points, however defined or accrued, or the conquest of some chunk of the playing area), but it—and one of these games—takes a long time to play. Since players often start from a single world, the game gets increasingly interesting and increasingly complex as they expand to other worlds and eventually contact the other players. Short games—time-limit games—thus tend to be wholly unsatisfactory, and the bookkeeping demands, particularly in the latter stages, can be forbidding and unforgiving of mistakes. Within their limits, despite these problems, these games are among the most challenging and enthralling of all wargames.

Evaluations

After the Holocaust (1977)

PUBLISHER: Simulations Publications, Inc.

SUGGESTED RETAIL PRICE: $14 (boxed)

SUBJECT: Although military conflict is possible, this is primarily an economic simulation set in America twenty years after an atomic war.

PLAYING TIME: At least five to six hours are needed for the complete game.

SCALE: This ranges between strategic and grand-strategic simulation. Military units are divisions. Each hex represents 190 kilometers, and each turn is a year.

SIZE: The 22″ x 34″ mounted mapboard uses 19mm (¾″) hexes.

BALANCE: The midwestern and southwestern regions are favored.

KEY FEATURES: This is basically a four-player game with provisions for one, two, or three, and is a wargame in only the broadest of terms. The system breaks down the national economic situation into five "sectors." Food, fuel, and metal stocks are used to feed people and produce goods, while transportation and industry are abstract measures of the capacity of the region. Labor (people) and money are the other major resources in the game. Each player is cast in the role of overseer of the welfare of an expanding regional quasi-nation and must allocate labor and mechanization to provide for the present and future needs of his region. New areas, annexed through political activities centering on the expenditure of money, provide new resources and labor to expand the production base. Trade is necessary for any region to realize its full potential; military actions are an expensive diversion. The map is used almost exclusively for visual representation of holdings and for calculating trade route distances, while most of the counters are used in record-keeping on sector tracks situated on the board.

COMMENTS: *After the Holocaust* employs a complicated system of simple activities to represent the economic workings of a nation. Starving people, unemployment, and depreciation prove very real and troublesome hurdles to economic growth. The irregular distribution of regional resources results in large stockpiles in some sectors and acute shortages in others. Trade is the name of the game, as a region like the Northeast or (less realistically) the Far West must soon realize—or shortly starve in ignorance. The same necessity makes the game unsuited to solitaire play. Those who presume gamers are a bunch of warmongers would be shocked by this game; military efforts tend to be mutually destructive and as such are best relegated to last-gasp efforts. Although considerably off the beaten path, this is of interest —even if it's not as realistic as it would have you believe.

EVALUATION:
Presentation—Very Good
Rules—Very Good
Playability—Good
Realism—Fair to Good
Complexity—8

OVERALL EVALUATION: Good

Alpha Omega (1977)

PUBLISHER: Battleline Publications

SUGGESTED RETAIL PRICE: $13 (boxed)

SUBJECT: Combat in outer space in the third millennium (A.D.).

PLAYING TIME: Thirty minutes to more than four hours are needed, depending on the scenario.

SCALE: This is a tactical-level simulation; counters represent individual spacecraft, support weapons, fighter groups, or celestial bodies. Each hex represents 186,000 miles (one light-second), and each turn spans six seconds.

SIZE: Two 20″ x 26″ boards form a 40″ x 26″ playing surface. The very large hexes are 32mm (1¼″) across.

BALANCE: In all scenarios, the side with the larger ships is favored.

KEY FEATURES: *Alpha Omega* uses an allotment system similar to those of other tactical space wargames and not unlike that of *StarSoldier*. Each ship has a limited energy supply with which to energize its weapons, drive, and defenses each turn.

Written allotment plots are required. Weapons systems, some of which vary from race to race, include beams, drive-powered weapons (Mason fields and dispersion fields), and such options as Xenolite bombs, energy webs, minefields, scramblers, and others—all of which have independent and quite varied uses and effects. (Not all are adequately explained.) Combat is by ranged fire; the proper column of the Combat Results Table is determined by various modifications, *not* the Attack/Defense odds. Something of a simplified vector system is used for movement, and hidden movement is an option.

COMMENTS: It's obvious from the first that the basic system design of *Alpha Omega* is a good one. What becomes obvious with experience is that the game wasn't adequately developed. By allocating a large amount of energy to its defensive-cloaking function, a very large ship can render itself virtually invulnerable to attack from almost any number of smaller vessels—which have no advantages at all. Even the optional weapons are insufficient to overcome this difficulty. The best scenario may be the first (solitaire) one; most of the others seem not to have been play-tested thoroughly—if at all. And all ships suffer from a poverty of energy; they can't utilize half their options. The Drove ships are the best off in that regard, but they have their own problems with movement. If you can find or construct a scenario with relatively equivalent forces, however, it will provide an entertaining challenge.

EVALUATION:
Presentation—Good to Very Good
Rules—Good
Playability—Good
Realism—Fair
Complexity—7

OVERALL EVALUATION: Good

BattleFleet Mars (1977)

PUBLISHER: Simulations Publications, Inc.

SUGGESTED RETAIL PRICE: $20 (boxed—less in other packaging)

SUBJECT: Interplanetary combat between the Earth-based Ares Corporation and a rebelling Martian colony, A.D. 2094.

PLAYING TIME: The game takes five to more than ten hours, depending on whether the tactical module is used (among other factors).

SCALE: This is a strategic simulation with a separate tactical-combat subgame. Units are individual ships and political agents. Turns represent one month.

SIZE: The strategic (main) game map is 22″ x 34″ and does not use hexes. The tactical display, if used, is a separate 22″ x 34″ map.

BALANCE: The strategic game favors the Ares Corporation.

KEY FEATURES: The strategic game is one of morale and attrition; players attempt to wear down each other by diminishing resources and morale points. Victory in any scenario is based on points for possession of territory or enemy forces destroyed. Movement is abstract but a very accurate representation of rocket-based interplanetary travel. There are provisions for sabotage, production, and revolutions. There are two tactical modules. The first is a quick and simple die-rolling abstraction. The second is a simulation of three-dimensional combat with full, realistic effects of Newtonian physics. Unfortunately, this requires a twinned display in which each ship is represented by five counters. Combat is by missile fire or high-powered lasers. It works better than it has any right to work, but it takes some getting used to. The full tactical game adds considerably to the playing time of the game.

COMMENTS: This is probably the best simulation of a war in space yet published, but despite its realism it doesn't work as a game. Although the rationale is well developed, there's too much complexity in the game for the interest it can sustain. Play is usually stopped by boredom, rather than by any conclusion. Victory is entirely too chancy, anyway; given the premises, this may be "realistic," but it's unsatisfying after a ten-hour game. A lot of effort has gone into *BattleFleet Mars*, but playing it takes more work than the subject justifies.

EVALUATION:
- Presentation—Very Good
- Rules—Good
- Playability—Fair
- Realism—Very Good
- Complexity—8

OVERALL EVALUATION: Fair

Black Hole (1978)

PUBLISHER: Metagaming

SUGGESTED RETAIL PRICE: $2.95 (plastic envelope)

SUBJECT: Combat on the surface of a toroidal (doughnut-shaped) asteroid.

PLAYING TIME: Thirty minutes to two hours are needed.

SCALE: This is a tactical game. Units are individual fighting vehicles, and turns represent one minute.

SIZE: The 8¼" x 21" game map uses 15mm (⅝") hexes.

BALANCE: This is even.

KEY FEATURES: The sequence of play is not a very complicated version of the standard Move/Shoot pattern. The forces available to each player consist of vehicles armed with missiles or lasers in a mix of the player's choice. The central feature of the game is the geography: units leaving one edge of the mapboard "reenter" on the opposite edge. Missiles that miss their target continue to orbit until they hit something. Lasers, being line-of-sight weapons, can fire at anything on the inner surface; it's all in view. Units can "land," and some can try to jump across the "hole," but both are risky: the black hole in the center may swallow them up.

COMMENTS: Other than the toroidal surface, this is just another wargame. The science is poor. Both asteroid and black hole (of the size and shape described) are shaky propositions, at best. The range modifications for laser effectiveness can't be justified technologically. Missiles should travel much faster than they do and, consequently, should not orbit. Nor is it realistic to have so little control over them; missile combat wasn't so haphazard ten years ago. Nonetheless, the artificial limitations and the somewhat warped science combine with the unique map to make an enjoyable, nonserious game, at least until the novelty wears off.

EVALUATION:
- Presentation—Fair
- Rules—Fair to Good
- Playability—Excellent
- Realism—Poor
- Complexity—5

OVERALL EVALUATION: Good, for a while

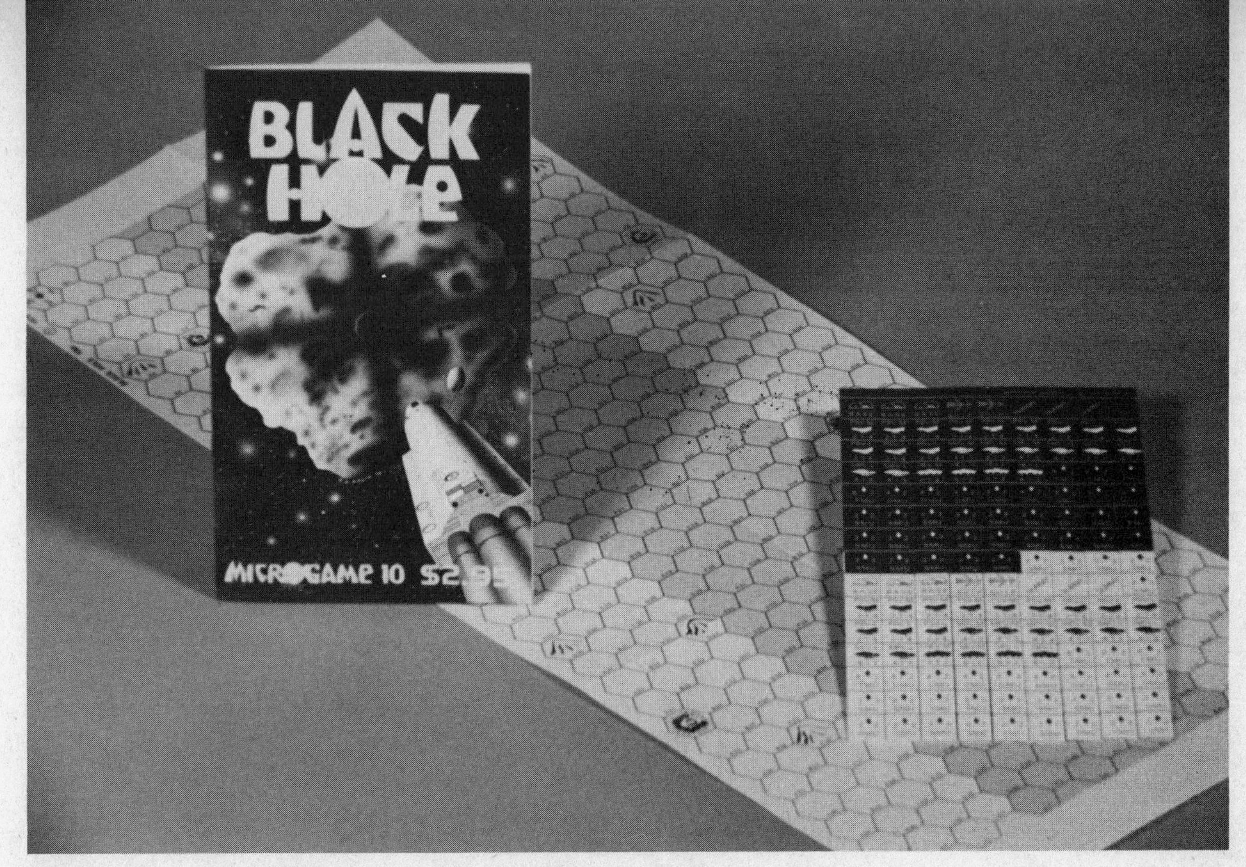

Metagaming's MicroGame Black Hole. (PHOTO: METAGAMING CONCEPTS)

4000 A.D. (1972)

PUBLISHER: House of Games, Inc.

SUGGESTED RETAIL PRICE: $14.95 (boxed)

SUBJECT: Interstellar expansion and conquest circa A.D. 4000.

PLAYING TIME: Two to four hours are needed, although four contemplative players might drag it out somewhat longer.

SCALE: This is a strategic game. Units (pieces) are individual starships. Time and distance scales are highly abstracted.

SIZE: Two 14″ x 21″ boards form a 28″ x 21″ playing surface. Squares are 5¼″ across; there are no hexes.

BALANCE: Quite good, although the "middle" player in the three-person version is at a slight disadvantage, and the turn order in all versions confers benefits on the player(s) going first.

KEY FEATURES: This is the simplest sort of exploration-and-conquest game: no hexes, no counters (plastic spaceships instead), no combat or movement factors, no Combat Results Table, and no "exploring." Stars are colored to give a vertical dimension to the board and coded to show their resources, an opposing pair of which is enough to build a new spaceship every other turn (rather like *Diplomacy*). Movement is unique. A "space warp" allows "hidden" movement in plain sight of everyone by showing only the time and distance of a fleet from its point of departure—not the destination. Ships may leave the warp on any turn and emerge at any appropriately distant (legal) star system. Combat is even simpler than *Diplomacy*'s. The greater number of ships at a star annihilates the lesser number without loss to the larger force. This makes large stationary forces as much of a hazard as a threat and traveling fleets a terror.

COMMENTS: This is a tense contest for those who like their games devoid of chance—but not of surprises. *4000 A.D.* is one of those rare games that are almost equally attractive for two, three, or four players. It's handsomely done, but at night the board is a bit too dark, and distinguishing the red ships from the red-orange ships would induce eyestrain in an eagle. (Hint: Paint the tips of one set or the other.) The alliance rules have a few holes—particularly with regard to fleet rendezvous—but these and the effects of turn order can be solved (at a slight bookkeeping cost) by using written orders and simultaneous moves. *4000 A.D.* is not only fun but proof positive that a game doesn't have to be wildly complicated to be a real challenge.

EVALUATION:

Presentation—Very Good
Rules—Good
Playability—Excellent
Realism—Poor
Complexity—2

OVERALL EVALUATION: Very Good

G. E. V. (1978)

PUBLISHER: Metagaming

SUGGESTED RETAIL PRICE: $3.95 (plastic envelope)

SUBJECT: Armor and infantry actions in the twenty-first-century world of Ogre.

PLAYING TIME: The game takes one to two hours.

SCALE: This is a tactical simulation. As in the game of *Ogre*, units represent infantry squads or individual vehicles or guns; hexes are six hundred meters across; turns last two to three minutes.

SIZE: The 12″ x 14″ map uses 15mm (⅝″) hexes.

BALANCE: This is very good.

KEY FEATURES: This is really an expansion kit for *Ogre*. To the craters and rubble in *Ogre* the new map adds woods, city, and swamp hexes as well as a railroad. There are some new units, including a train in one scenario, new stacking rules, and a

Metagaming's MicroGame G. E. V. (PHOTO: METAGAMING CONCEPTS)

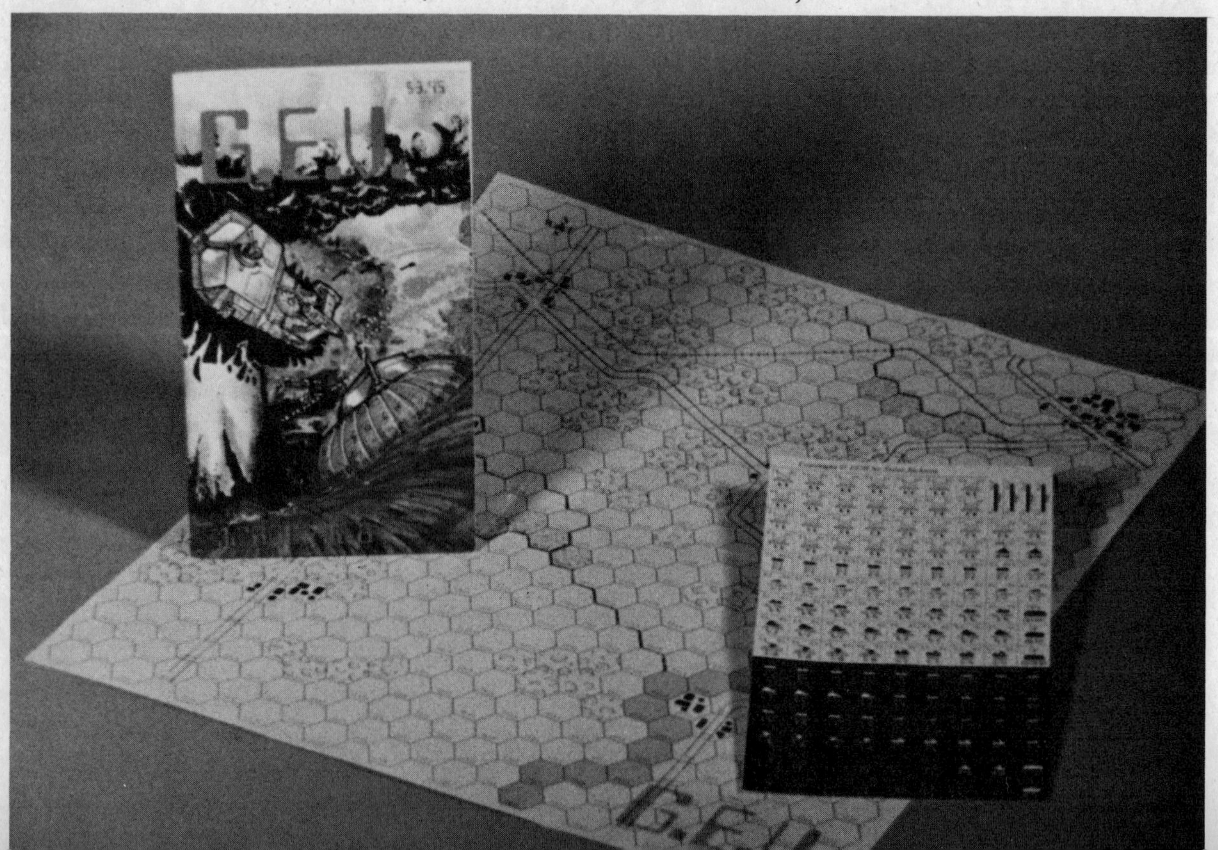

considerably more complicated overrun procedure. The objective of this game is not defense against an Ogre but the destruction of enemy units, so the strategic problems are much more like those of a conventional wargame.

COMMENTS: *G. E. V.* suffers from the problem of all sequels: it's hard to live up to the reputation created by the original. Since *Ogre* is one of the most remarkably successful games ever, *G. E. V.* has a hard act to follow. It can perhaps be forgiven its small failures. As a game in its own right, it's somewhat dull. Though the situation is hypothetical, the feel is exceptionally conventional. The battles could almost be lifted whole from *PanzerBlitz*. Only when Ogres are added and the situation is reduced to the uncomplicated objectives of the earlier game does it come to life. It's a good expansion kit, but only a fair game. If you have *Ogre* and like it, you'll want this. If not, get *Ogre* first.

EVALUATION:
Presentation—Fair to Good
Rules—Good to Very Good
Playability—Very Good
Realism—Very Good
Complexity—6

OVERALL EVALUATION: Good

Holy War (1979)

PUBLISHER: Metagaming

SUGGESTED RETAIL PRICE: $2.95 (plastic envelope)

SUBJECT: An interstellar "religious" war between inhabitants of a "pocket universe" located *within* a giant creature.

PLAYING TIME: About one hour is needed.

SCALE: This is a strategic game with tactical overtones. The hexes represent one light-year (sort of), and each turn represents one day. Units are huge individual ships and starguard forces.

SIZE: The 12″ x 14″ board uses large hexes that are almost 44mm (1¾″) across. Within each hex, there are seven smaller hexes.

BALANCE: A slight edge is given to the Holy Band (strategically, the defenders).

KEY FEATURES: By means of a clever map display that uses a spiral chain of progressively smaller (and thus "lower") hexes within each large hex, true three-dimensional movement is allowed. Also available is movement by "warp lines," which is much faster but subject to the availability of an access point. The selection of ship types is exemplary (if somewhat confusing); it includes ships that can create such warp lines on a limited basis, shove stars around, or convert adversaries. There are some interesting provisions of combat—three combat factors, two major types of combat and corresponding Combat Results Tables (one differential and one odds/ratio), random selection of weapons, and so on—all very idiosyncratic and not entirely welcome.

COMMENTS: The rationale for the confusingly named *Holy War* is quite complex and peculiar. It should please those who believe science-fiction games should be unconventional. As a direct and obvious consequence, of course, it suffers the same disadvantages as the original fantasy games of the next chapter: the thing is bewildering. It apparently began life as a full-sized game, and the transition to the microgame format was not entirely successful. This isn't a bad game, but there's too much of it. If you drop about one-third of the rules, the result is a reasonably

Metagaming's MicroGame Holy War. (PHOTO: METAGAMING CONCEPTS)

clean and simple game less dependent on luck than the original. For three dollars, you can afford to do whatever you want.

EVALUATION:
Presentation—Good
Rules—Good
Playability—Fair
Realism—Fair
Complexity—7

OVERALL EVALUATION: Fair to Good

Imperium (1977)

PUBLISHER: Conflict Game Co.

SUGGESTED RETAIL PRICE: $9.98 (boxed)

SUBJECT: Interstellar warfare in the distant future between a Terran confederation and a galactic empire (the Imperium).

PLAYING TIME: Two to five hours are needed per game; several times as long for a complete campaign.

SCALE: This is a strategic game with tactical overtones. Hexes are half a parsec (more than 1½ light-years) across; turns equal two years. Counters represent squadrons of various kinds of ships, unspecified (but large) contingents of troops, outposts, and so on.

SIZE: The 19″ x 25″ game board uses 16mm (⅝″) hexes.

BALANCE: Although sympathies lie with the Terrans, the asymmetrical ship designs and starting positions give the Imperium a bit of an edge.

KEY FEATURES: Each player can choose from a wide variety of units—from ships to colonies—whose cost (in time and money) and capabilities also vary greatly. The Terran player gets income from bases and colonies; he must build to expand and must expand to build. While the Imperial player is also affected by these factors, most of his income is derived from the Imperial budget (and hence is relatively fixed). He may appeal to the Emperor for a budget increase or build the bigger (and otherwise prohibited) kinds of warships. A sliding scale of Glory Points serves the threefold function of providing a cost for such appeals, the conditions for victory, and a time limit for the game. Units have no movement factors. Movement is instantaneous over "jump lines" and hopelessly slow otherwise. Aside from making control of jump lines important, this effectively puts terrain on the board and obviates the necessity for a three-dimensional display or movement system —most of which don't work. Considering the scale, combat is surprisingly tactical in flavor. There are two different weapons systems (beams and missiles)—present in varying degrees on different ship types—each of which is effective at one of the two ranges at which combat takes place. Planetary assault is also possible. The Combat Results Tables integrate attack and defense factors (but not as an odds/ratio) and are decidedly lethal. The sequence of play is clear but quite involved for the overall complexity level of the game.

COMMENTS: A lot of thought went into *Imperium*. The presentation is flawless. The silhouette counters are colorful and clearly read; the board is stark but striking; the Fabian cover is right out of *Star Wars;* all of the charts and tables are on a pair (one per player) of page-size cards. The rules can be understood. While not wildly original, the rationale is suitable and appealing. Various tables add "historical" flavor and the tang of chance without controlling the outcome of the game. At both the strategic and tactical levels, players have almost a surfeit of options. The attractions of a long campaign are present without the usual disadvantages; as with *Richthofen's War,* it can be played a game at a time and put away between sessions. It's marvelous fun and challenging, too. Hard to beat.

EVALUATION:
- Presentation—Excellent
- Rules—Very Good
- Playability—Very Good
- Realism—Good
- Complexity—6

OVERALL EVALUATION: Very Good

Invasion: America (1975)

PUBLISHER: Simulations Publications, Inc.

SUGGESTED RETAIL PRICE: $20 (boxed, with mounted board) or $14 (plastic box, with unmounted board)

SUBJECT: Conventional land combat during a hypothetical invasion of North America in 1997.

PLAYING TIME: Three to ten hours are needed, depending on the scenario and the number of players involved.

SCALE: This is a strategic-level simulation; units represent divisions or corps. Each hex represents 130 kilometers, and each turn spans one month.

SIZE: Two 22" x 34" maps overlap to form a 35" x 42" playing surface. Hexes are 16mm (5/8") across.

BALANCE: Six scenarios accommodate two to five players; some scenarios pose greater difficulties for the U.S.-Canadian player than do others.

KEY FEATURES: *Invasion: America* reveals a

simple yet effective system for staging amphibious invasions. Rules for air and naval units are abstracted (and prove the defender's real advantage). The U.S.-Canadian player may also exercise limited strategic railroad movement. Special units represent hovercraft, militia, and Asian "wave infantry" troops, and special rules detail partisan activity, the U.S. annexation of Mexico, and the rigors of weather in northern areas. The key innovation here is the use of "untried" units, whose strengths are known only approximately until they have been involved in combat.

COMMENTS: To the basic gimmick of untried units is added the rare attraction of a modern battle fought in North America. Unusual in a game with multiplayer potential, *Invasion: America*'s solitaire playability is high. The graphics of the mapboard were also an advance in the state of the art for SPI. There is some question concerning the victory conditions for some of the scenarios (the point totals required for victory are unattainable unless urban supply hexes are counted as eight victory points apiece, which is intended but not specifically stated). Regardless of the scenario, each side has unique advantages and a wealth of strategic options. This variety and the game's progressive design make *Invasion: America* a welcome addition to the library of anyone interested in modern warfare.

EVALUATION:
Presentation—Very Good
Rules—Good
Playability—Very Good
Realism—Good
Complexity—6

OVERALL EVALUATION: Very Good

Invasion: America, *an SPI simulation set in the year 1997.* (PHOTO: SIMULATIONS PUBLICATIONS, INC.)

Ogre (1977)

PUBLISHER: Metagaming

SUGGESTED RETAIL PRICE: $2.95 (plastic envelope)

SUBJECT: Twenty-first-century armor and infantry actions involving a massive, cybernetic tank (the "Ogre").

PLAYING TIME: Less than one hour is required.

SCALE: In this tactical simulation, units represent infantry squads or individual vehicles or guns. Hexes are six hundred meters across; turns represent roughly two to three minutes.

SIZE: The 8″ x 14″ map uses 15mm (⅝″) hexes.

BALANCE: This is very good.

KEY FEATURES: The situation in basic *Ogre* involves an assault against a prepared command position by a single supertank.

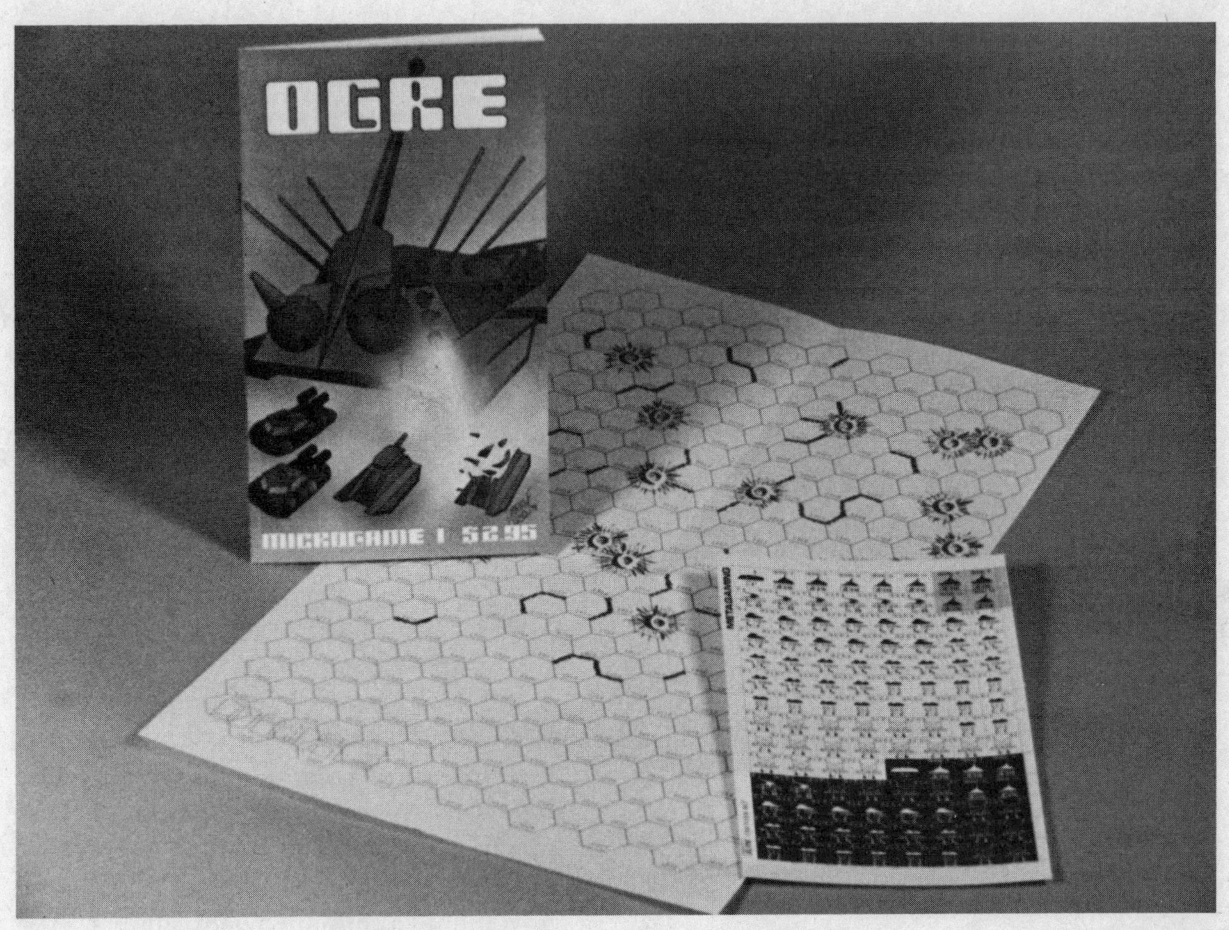

Metagaming's MicroGame Ogre. (PHOTO: METAGAMING CONCEPTS)

Other versions of the game, which can be played with multiple commanders and as much complexity as you care to add, are variations on this theme. Combat is by ranged fire generally, but the Ogre can also execute overrun attacks. For practical purposes, the Ogre is treated almost like an inseparable stack of units: its weapons can be directed at different targets, and they and the treads that govern the Ogre's movement must be attacked (and destroyed) separately. Lesser opposing units —smaller tanks, GEV's, howitzers, and infantry squads—work in conventional tactical fashion. The very small number of units involved—a handful on one side and just one (the Ogre) on the other—reduces the effective complexity of what is a fairly sophisticated game system.

COMMENTS: *Ogre* is the game that made Metagaming's "MicroGames" one of the best commercial ideas in recent game marketing. It is ridiculously inexpensive—almost disposable. Despite its size and price, it is well produced and reasonably presented. The map is thick enough to last; the counters are printed on kraft stock and, while cut in only one direction, are easily separated with scissors. Most significantly, perhaps, it's an exceptionally fast and interesting game. The situation is classic. Any reasonable attack the Ogre makes will have a fifty percent chance of destroying its target, but there are always more targets available than it has weapons. As the defender, you cannot prevent the Ogre from munching away at your forces, but you must stop it from gulping you down whole. A pretty dilemma. Its drawbacks are minor. Despite its claims, it's really no more suitable for solitaire play than most others. Since it is usually easy to find an opponent for so short and playable a game, however, this isn't a problem. It's one of the best values in gaming.

EVALUATION:
Presentation—Fair
Rules—Excellent
Playability—Excellent
Realism—Good
Complexity—4

OVERALL EVALUATION: Excellent

Outreach (1976)

PUBLISHER: Simulations Publications, Inc.

SUGGESTED RETAIL PRICE: $10 (plastic box; also available with *StarForce* and *StarSoldier* as *StarForce Trilogy* for $24)

SUBJECT: Exploration, colonization, and conflict in our galaxy, circa A.D. 3000.

PLAYING TIME: As little as three hours are needed for a short scenario, but with three or four players, ten to fifteen hours is more typical.

SCALE: *Outreach* is a grand-strategic game. Units represent regular, explorer, or dreadnought starfleets. A hex is twelve hundred light-years by fifteen hundred light-years. A turn represents the passing of a generation.

SIZE: The 22″ x 34″ game map uses 19mm (¾″) hexes.

BALANCE: There are five scenarios for one or more players; some pose definite positional disadvantages for various players.

KEY FEATURES: Except for the action of periodically discovered, "autonomous" (nonplayer) alien races, the game system is simpler than most games of this kind. The focus is always on the construction and maintenance of a chain of stargates, which is essential to movement even when it is not a condition of victory. The Fate Table of random events has profound effects on play, but a die roll on it may be modified in various ways at player discretion. In some scenarios, an increase in civilization level—possible only by rolling on the Fate Table—is required for victory.

COMMENTS: The board is striking—almost dazzling—and *Outreach* is fun to play. *But*. It takes too long, particularly the multiplayer scenarios, which are potentially the most enjoyable. If the initial movement allotment were higher, the game would develop faster and be more like World War II than World War I—but it isn't. Combat is prohibitively expensive —properly and realistically so, perhaps, but there would be more spice to the game if conflict were not limited to maneuvering for positional advantage. As in nearly all games of this general sort, there is considerable bookkeeping involved, and it's easy to make errors that can get so compounded the final outcome is meaningless for all concerned. There are not enough alien systems or counters. You cannot modify the Fate Table out of existence, and the wrong roll can easily take a player completely out of contention. The chance factor is just too high. Despite all that and some flat contradictions in the rules, this game can be a pleasant way to spend a Saturday—if you don't take your gaming too seriously.

EVALUATION:
Presentation—Very Good
Rules—Good
Playability—Fair to Good
Realism—Fair to Good
Complexity—6

OVERALL EVALUATION: Good

StarForce (1974)

PUBLISHER: Simulations Publications, Inc.

SUGGESTED RETAIL PRICE: $10 (plastic box; also available with *Outreach* and *StarSoldier* as *StarForce Trilogy* for $24)

SUBJECT: Interstellar warfare in the twenty-fifth century.

PLAYING TIME: It takes two to more than eight hours, depending on the scenario and whether the basic or advanced combat system is being used.

SCALE: On the strategic board, each hex represents one light-year. On the tactical display (in the Advanced Game), one hex is equal to one-third of a light-day. Units represent starforces ("fleets" of four ships) or stargates.

SIZE: The 22″ x 34″ unmounted mapsheet uses 16mm (5⁄8″) hexes.

BALANCE: Fifteen scenarios offer plenty of choice.

KEY FEATURES: While victory is attained by military conquest of the opponent's stargates, this is essentially a game of maneuver. In fact, the importance of the stargates is due to their ability to increase the range and accuracy of starforce movement. Since units teleport ("shift"), only the simplest sort of simultaneous movement is required in the Basic Game: only the starting and ending hexes are written. The addition in the Advanced Game of a system for tactical combat and movement—and *those* written plots—reduces playability (and lengthens playing time) considerably. Both forms of combat use differential Combat Results Tables that only affect the defender. There are scenarios for one to three players, and some sketchy suggestions for verbal-plotting techniques are included.

COMMENTS: The long playing times caused by written movement plots are further extended by the Advanced Game's combat system. Combined with the use of "faker forces" (starships masquerading as starforces; effectively, dummy counters),

StarForce, *an SPI game of interstellar conflict in the twenty-fifth century.* (PHOTO: SIMULATIONS PUBLICATIONS, INC.)

however, the limited intelligence arising from the simultaneous movement system creates an air of tension seldom realized in wargames. While combat centers around the stargates, a player has little knowledge from which to extrapolate threats and weaknesses on the part of his opponent. A patient gamer will be rewarded with a tense contest that genuinely captures a hint of the unknown; the feel of this game is distinctly unconventional—sometimes uncomfortably so. For the impatient, some scenarios require fewer counters than others.

EVALUATION:
Presentation—Very Good
Rules—Very Good
Playability—Fair
Realism—Good
Complexity—7

OVERALL EVALUATION: Good

Starship Troopers (1976)

PUBLISHER: Avalon Hill Game Co.

SUGGESTED RETAIL PRICE: $12 (boxed)

SUBJECT: *Starship Troopers* is a simulation of land combat on another planet in the twenty-second century between humans and two alien races. The game is adapted from Robert Heinlein's novel of the same name.

PLAYING TIME: It takes two to four hours.

SCALE: The game is a close-tactical simulation. Each counter represents one soldier or weapons platform, and each hex represents a distance of one mile.

SIZE: The 24″ x 22″ game board uses 16mm (5/8″) hexes.

BALANCE: Eight scenarios provide equal opportunities for each side.

KEY FEATURES: *Starship Troopers* uses a standard Move/Fire/Melee sequence with the addition of a special-functions phase at the beginning and a second-movement phase for the human player after close combat (melee). On separate Combat Results Tables, the humans suffer stuns, various wounds, or death, while the aliens are liable to disruption or elimination. Special counters include heavy weapons units, air cars, command units, scouts, and others. Inverted counters and hidden-tunnel complexes designed by the "Bug" player are used to create the effect of limited intelligence. These provisions make the game less than ideal for solitaire play, but suggestions are included to aid the solo gamer. The rules were among the first to employ the programmed sequential learning approach, which helps the gamer adapt more easily to the peculiarities of the situation (and the system) by adding complications gradually, scenario by scenario. This makes the game easier to play than its ultimate complexity would indicate.

COMMENTS: *Starship Troopers* provides a refreshingly different environment in which to maneuver a fairly standard array of units. Against the arachnids, the human player must adapt to truly alien conditions to discover and eventually root out the Bugs from their subterranean tunnel complexes. Against the humanoids, more conventional tactics may be employed, and the game resembles an airdrop from any other era. If the game has a fault, it may be that it's *too* conventional: it's *PanzerBlitz* in spacesuits. But the action is fast and furious, and the game is as tense and exciting as any on the market. Despite some minor exceptions, the game is relatively faithful to the book—which

Starship Troopers, a game based on Robert Heinlein's book. (PHOTO: AVALON HILL GAME CO.)

was good enough to win a Hugo, science fiction's version of the Oscar.

EVALUATION:
Presentation—Very Good
Rules—Excellent
Playability—Very Good
Realism—Very Good
Complexity—6

OVERALL EVALUATION: Very Good

StarSoldier (1977)

PUBLISHER: Simulations Publications, Inc.

SUGGESTED RETAIL PRICE: $10 (plastic box)

SUBJECT: Land combat in the twenty-fifth century.

PLAYING TIME: It takes three to five hours.

SCALE: This is a close-tactical (man-to-man) simulation; each hex represents a kilometer.

SIZE: The 22″ x 34″ game map uses 16mm (5/8″) hexes. Room is also required for a 8½″ x 11″ soldier-status display for each player.

BALANCE: Variable situations and orders of battle make possible more than a hundred scenarios and allow players to adjust balance as necessary. However, in many scenarios one player has orbital-bombardment capabilities, which can unbalance the game in his favor.

KEY FEATURES: The underlying system is more commonly found in tactical naval games. Units have a limited (and racially varied) task-point allowance for each turn. These task points are spread out among such functions as movement, direct fire, launching projectiles, and defensive countermeasures. Combat can result in lost task points that can be recovered in time by soldiers fortunate enough to remain unmolested for a complete turn. Outright death is infrequent. Android troops and even a race with a communal mind (the Rame) are represented. (We suggest that Rame bases not be allowed to combine with or benefit from the countermeasures of Rame starsoldiers in the same hex. Otherwise they tend to be indestructible.)

COMMENTS: *StarSoldier* is very similar in concept to two other SPI man-to-man tactical games, *Sniper!* and *Patrol.* While the latter games are set in the twentieth century and have no task-point allotment as such, all of them tend to emphasize the importance of movement and terrain to defense; all three utilize written movement plots and opportunity fire; all have the same personal feel. In *StarSoldier* the written plot includes all functions performed by each unit each turn, but the small number of units in play makes this bearable—barely—unless you're just allergic to simultaneous movement. Games are very fluid as units flit from cover to cover, but most boil down to putting out an overwhelming offense while maintaining an adequate defense—a combination few units can manage. The victory conditions are designed around the probable lethality of the future soldier, and the game is usually a race to succeed before one side is devastated to the man. *StarSoldier* is the smallest scale game in a trilogy that includes *Outreach* and *StarForce.*

EVALUATION:
Presentation—Good
Rules—Good
Playability—Good
Realism—Good
Complexity—7

OVERALL EVALUATION: Good

WarpWar (1977)

PUBLISHER: Metagaming Concepts

SUGGESTED RETAIL PRICE: $2.95

SUBJECT: Interstellar war in the distant future.

PLAYING TIME: An hour or two is needed.

SCALE: This is a strategic game with some tactical overtones. Units are individual ships. Time and distance scales are highly abstracted.

SIZE: The 8″ x 14″ game map uses 15mm (5/8″) hexes.

BALANCE: This is dead even.

KEY FEATURES: The ultimate object is the occupation of enemy base stars, but the focus is on the control of warp lines, which allow cheap movement from star to star, and do-it-yourself ship designing. Each player gets an allotment of Build Points (BPs) with which to build whatever ships he pleases. Most systems have a variable cost in BPs: the more points built into them, the more powerful they are. Power/Drive powers the other systems and is the ship's effective movement factor (this is similar to *StarSoldier, Alpha*

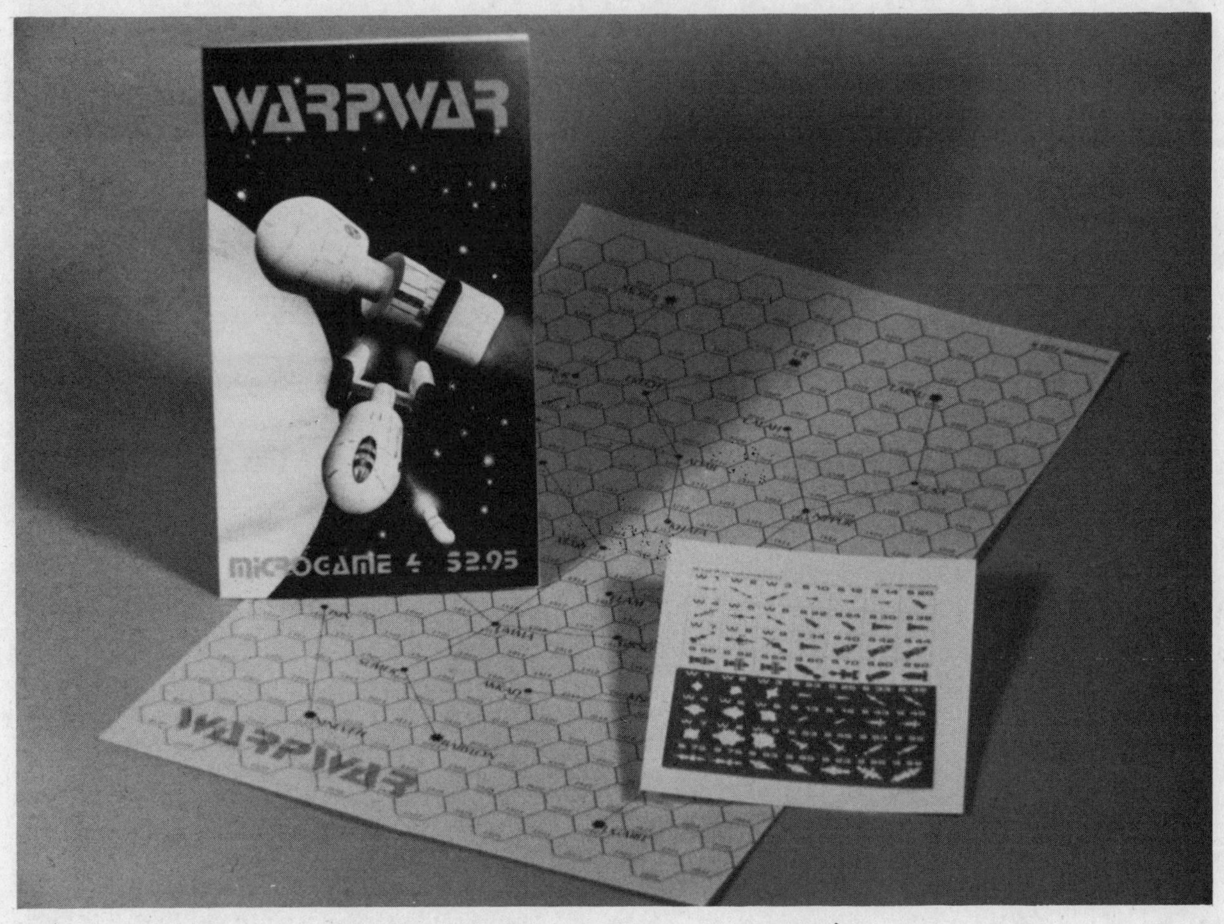

Metagaming's Microgame WarpWar. (PHOTO: METAGAMING CONCEPTS)

Omega, and, for that matter, *Starfleet Orion*). Beams and Missiles are the offensive weapons; Screens provide some defense. Systemship Racks allow Warships (those with Warp Generators) to carry Systemships (those without), which, for practical purposes, can fight but cannot move. Combat is of the "nonrandom" sort and depends on the interaction of the options (three) chosen by each ship and the difference between their two drives in the combat round. Damage, if any, is the difference between the points applied to the Beam (or, for a Missile, just an arbitrary two points) of the attacker and the Screen of the defender—aided and abetted, in either case, by the ship's technology level.

COMMENTS: Despite the fact that ships are the only units in the game, this is less a tactical space combat game like *Alpha Omega* (or, especially, *Starfleet Orion*) than it is a simple country cousin of *Imperium* or *StarForce*. Individually or collectively, there's not much new here except another try at a diceless Combat Results Table. This one is not so obnoxious as some, but since the options are unequal, combat gets tedious, and play can become somewhat stereotyped at both the tactical and strategic levels. There are a variety of nits to pick. Nineveh, Babylon, Ugarit, and the rest look at little out of place on a star map, and the rules don't seem to permit a victory in the Advanced Game. For a small, short game, there is a great deal of (admittedly uncomplicated) bookkeeping involved. As a whole, this is probably the common impression of microgames: they are not something you'd play for years, but can be short, quick fun for a while.

EVALUATION:

- Presentation—Fair
- Rules—Good
- Playability—Very Good
- Realism—Fair
- Complexity—4

OVERALL EVALUATION: Good

CHAPTER 15

Fantasy Games: Swords & Spells, Quest & Conquest

MOST PEOPLE know—or think they know—what science fiction is, but for some the term "fantasy" suggests either too much or too little to be helpful. To say, as the standard attempts to distinguish them do, that science fiction deals with the imaginary-but-possible, while fantasy covers the imaginary-and-impossible, begs the question of what is "impossible." Many people thought it impossible for the Wright brothers to get their contraption off the ground or for rockets to work in space without air to push against. There are scientists today who view the existence of advanced, intelligent life elsewhere in the universe as a statistical certainly—but believe it impossible for any UFO sighting to be evidence of the fact. One film reviewer found the devils of *The Exorcist* and *Rosemary's Baby* more believable than the creature in *Alien*. Dragging in intent may raise its own questions, but at least it allows us to proceed. The writer-creator of fantasy is deliberately working with material he believes is contrary to the way things really are.

Into the realm of fantasy we can put witches on broomsticks, vampires and werewolves, deals with the devil, Santa Claus (sorry, Virginia), the stork as an agent of human reproduction, talking mules, beagles in biplanes, Mickey Mouse, most monster movies, comic-book superheroes, Never-Never Land (sigh), and the movie *Heaven Can Wait*. However, the most commonly recognized sort—and virtually the only one impinging on wargames—is the subclass known as heroic fantasy or "sword and sorcery." This is sometimes described as "fairy tales for adults" and is exemplified by Beowulf and the dragon; Sir Gawain and the green knight; the hammer of Thor and the sword that was broken; the magic of Merlin, Circe, or Gandalf; the mighty thews of Conan, Fafhrd, and John Carter of Virginia and Mars; the worlds of Nehwon, Melnibone, Aquilonia, and, especially, the Middle Earth of J. R. R. Tolkien's *The Lord of the Rings*.

If these names don't quicken your pulse, evoke images of dark towers, oviparous maidens with reddish skins, and a silent flute whose song is heard only by the proper pilgrim, then this is going to be strange and puzzling territory indeed, and you are better off returning to firmer ground. This is no country for artillery, and an armored division is much too heavy for the ferryman's raft. Of course, if you've trod the glory road before, you know the fare; pay the old man, and we'll be on our way.

THUNDERBOLT AND LIGHTFOOT

The biggest problem of fantasy games comes from the very nature of fantasy. While game designers have (we hope) read enough military history to know why the French assault faltered at Agincourt or how effective chain mail was against an arquebus, it's not easy to find sources for the battleworthiness of unicorns and the typical sidearms carried by Ringwraiths. A complete order of battle for the siege of Barad-dûr is not immediately available, and the TOE of dwarfish cavalry is still a matter of much contention. While the designers of conventional games can solve almost all of their problems by studious research, this is only the beginning for the poor soul who hopes to do a fantasy game.

Furthermore, the basis of any wargame is quantification or the reduction of all relevant aspects of a conflict into numbers: movement factors, combat factors, range, distance. Yet fantasy is based on elements that do not quantify easily, factors that differ in quality—not quantity—from the norm. Among them are hand-to-hand combat, magic, and personality.

Tales of heroic fantasy usually take place in a setting reminiscent of medieval romance (somewhere between King Arthur's Camelot and Spenser's *Faerie Queene*) or an equally romanticized version of antediluvian civilization. In either case, the weaponry involved—swords, spears, shields, and armor—is "personal," and combat is heroic, bloody, and immediate. Depending on the intent and skill of the author, the battle scenes will either be true (if sometimes glorified) realizations of sword-and-shield fighting, or they will involve a lot of hewing about with mighty blades, harvesting enemies like standing wheat, and being generally as realistic as the standard cowboys-and-Indians clashes of old western B movies, whose heroes could fell ten Apaches with a single bullet.

Even if the combat is realistic, it has all the inherent difficulties discussed in Chapters 6 and 7: all the interesting and significant aspects of a battle take place on the personal level. (You'll recall that in the game *Iliad* and, to a lesser extent, *Troy*, combat in the game was strictly between individuals.) It is not strategy or tactics but prowess that counts. It's even harder to treat battles depending on "tigerish swings" and "pantherish moves," and it's next to impossible to balance a game in which one side—or one character—is invincible by auctorial decree.

Magic is an even thornier problem. In most fantasies, there is an element of the mystical, the supernatural, or the magical. In very few, however, are there any rules that govern the magic. Only because the author knows the limits his readers will accept, are the magicians, good and evil, prevented from settling the whole matter with a couple of quick thaumaturgic nuclear blasts. The limits on magic are rarely spelled out in a fashion a game designer can use to make a convincing, enjoyable, and playable game. Magic—more than submarines, more than helicopters—is the single item most difficult to simulate effectively in a wargame.

The attempts to include magic have been varied. Some games establish certain rules by which the wielders of magic are limited in what they can do. On occasion, designers have based these limitations on the rules followed by the alchemists and sorcerous experimenters of the Dark Ages. In other cases, magic has been treated as an effect of leadership or morale. In still others, it becomes just another type of catapult: Gandalf fires a thunderbolt and achieves 3–1 odds on a stack of Nazgûl—all are dispersed. Because the nature of fantasy demands—and gamers expect—an adequate treatment of this key element, the success of a fantasy game is often directly dependent on its success in handling magic.

The most memorable tales of fantasy are recalled as much for the personalities they contain as for the events of the story. We remember not just the slaying of the dragon but also the hero who slew it and the mark it made on his character. Fantasy board games can never hope to have the same success as the role-playing games of the next chapter at re-creating the element of personality. To *be* Saruman, Ningauble, or Conan in a game like *Dungeons & Dragons* requires only a natural exercise of the imagination. You just have to "act the part," as it were. To realize the presence of such a character in a board game—to personalize a cardboard counter that says "Saruman 4–6"—is much more difficult. The systems and mechanics are not suited to handling abstractions of personality, and there are no factors in the game that allow the players to make up for these deficiencies.

WHY CAN'T THE ENGLISH . . . ?

Fantasy board games can be divided into two classes: those based on some myth, epic, or work of fiction (like *Iliad* or *War of the Ring*) and those set in an original fantasy world devised by the game designers (*Sorcerer*, *Swords & Sorcery*). Both have their own special difficulties.

In the case of an adaptation, not only must essentially nonquantifiable aspects like magic and personality be reduced to the mundane systems and numbers of a wargame, but something like the total reality of the world from which the game is drawn must be preserved. There are plenty of gamers sufficiently conversant with the warfare of the Middle Ages to become upset if the power of the longbow is not properly indicated in the game. There are also gamers who will be displeased if the hero's magic sword, Glirendree, does not fell trolls left and right *and* protect all in its aegis from the fire of the archers. Aragorn ought to be immune to the lure of the One Ring, but Boromir should be tempted to seize it for himself. Since many of the necessary factors are subtle and involved, the translation poses many difficulties—and, of course, there will always be a vocal group who will be unhappy about how any number of incidents or characters are handled.

That's the *easy* part, though. The hard part is reconstructing a plot in which, if the author knows what he is doing, all things point to a single end—remaining faithful to that plot—while simultaneously allowing different (perhaps radically different) events and endings. How can you simulate the Trojan horse or the quest to destroy the ring in the fires of Mount Doom when the success of those ventures depended entirely on surprise? If Frodo and Sam give the ring to Gandalf to fight it out with Sauron, while they go back to the Shire to raid Farmer Maggot's fields and fish on the Brandywine, purists will howl. On the other hand, if Gandalf *can't* take the ring, and Frodo *must* bear it to the fires of Mount Doom—*and Sauron's side knows it*—the hobbits have no more chance of destroying the thing than the Children's Crusade had of retaking the Holy Land. There is no real escape from this double bind, and the only feasible "solution" is an uneasy compromise between event and possibility. The impossibility of a perfect solution does not, however, deter fantasy fans from clamoring for a board game version of their favorite tale—or designers from trying to concoct one.

The freedom offered by original fantasies appeals to some game designers, particularly those who don't know better. It looks easy. There's no historical research required, and no book to follow. It's the same attraction conventional writers find in literary science fiction and fantasy—generally with the same result. Without knowing the lay of the land, writer-designers on their first safari blaze "new" trails on old highways and return with "fish stories" everybody's heard before and no one would believe anyway. There are "rules"—accepted conventions—for the effects of silver and iron and the vulnerabilities of werewolves and dragons, and a designer flouts them at his peril.

Game designers who know fantasy and succeed in creating something truly novel —The Chaosium's *White Bear and Red Moon* or *Nomad Gods*, for instance—have a quite different problem to contend with. The world of Glorantha is strange even to readers of Tolkien and Howard or to students of ancient mythology. Venturing into genuinely foreign game territory requires hours spent not just learning the rules—the mechanics—but absorbing the setting, situation, and point of it all. Questions, ambiguities, and loopholes cannot be resolved through common sense or recourse to "the book," because there isn't one. The trip, therefore, requires an investment of time, energy, and imagination that is large even for wargames, and the incentive is not great enough to attract a large audience.

Despite their difficulties, fantasy games are more popular than ever, and, increasingly, the techniques of role-playing games are being applied to overcome the limitations of the board game format. They are not "simulations of reality" in any strict sense, but that is their charm. If the real world seems fresh out of damsels in distress and soft-hearted giant apes, they can still be found elsewhere. If quests are out of date and chivalry out of fashion, there is no Dragon Defense League to brand you a warmonger for slaughtering cardboard ogres. They won't appeal to every wargamer, but where else can you find out how Aragorn would have done against a Balrog?

Evaluations

The Creature That Ate Sheboygan (1979)

PUBLISHER: Simulations Publications, Inc.

SUGGESTED RETAIL PRICE: $3.95 (plastic envelope)

SUBJECT: An attack on a "typical" city by a creature from the monster movie of your choice.

PLAYING TIME: The game takes one to three hours, depending on the scenario.

SCALE: This is a tactical simulation, but except for the monster, the unit sizes and time scale are fairly abstract. One "square" is roughly equal to a city block.

SIZE: The 11″ x 17″ map uses a system of "squares" of varying sizes and shapes.

BALANCE: Without special powers, the creature is at a disadvantage in the learning scenario. Otherwise, balance is fairly even but depends heavily on the particular "mix" of powers chosen; the obvious balanced allotment is neither the most interesting nor the most effective. As far as interest goes, what red-blooded American would rather *not* play the monster?

KEY FEATURES: This game is "*Ogre* meets Creature Features": A single powerful unit (the monster) versus a swarm of lesser armor, artillery, infantry (National Guard, of course), police, and populace units. Unit selection is important for the human player; similarly, the monster "buys" factors for Attack, Defense, Building Destruction, and Movement—plus special powers like flying, web-spinning, and fire-breathing—which are reduced by successful human attacks. There are two Combat Results Tables, one for man-monster combat, another for building destruction—and an implied third one for starting fires. The monster tries to accumulate the necessary total of victory points by stomping buildings and chomping any of the populace too slow to get out of the way. The nameless monster counters include recognizable versions of Rodan, Godzilla, and other familiar types. Since only one monster exists in a normal scenario, the differentiation is purely for flavor—but flavor is what this game is all about. Several scenarios are included, but any number of others may be concocted.

COMMENTS: It seems appropriate that one of the first "capsule games," SPI's answer to Metagaming's MicroGames, should echo *Orge*, the game that brought back reasonable prices, brief rules, short games, and pure enjoyment. *The Creature That Ate Sheboygan* is inexpensive and eminently playable. The rules, however, while short and generally comprehensible (and relatively free of standard SPI legalese), contain several important chicken-or-the-egg ambiguities about fires. The bridge rules would be clearer if the bridge squares were outlined in red like everything else, and there seem to be some other red lines missing from two large buildings. Nor, despite the label, is this "science fiction" except in the loosest sense. No matter. If you don't find the basic notion irresistible, you've never watched Godzilla ravage Tokyo for the tenth time—and you wouldn't know what to do with the game anyhow. It really is nearly as much fun as it appears, particularly if your monster is equipped with fire-breathing. Closet py-

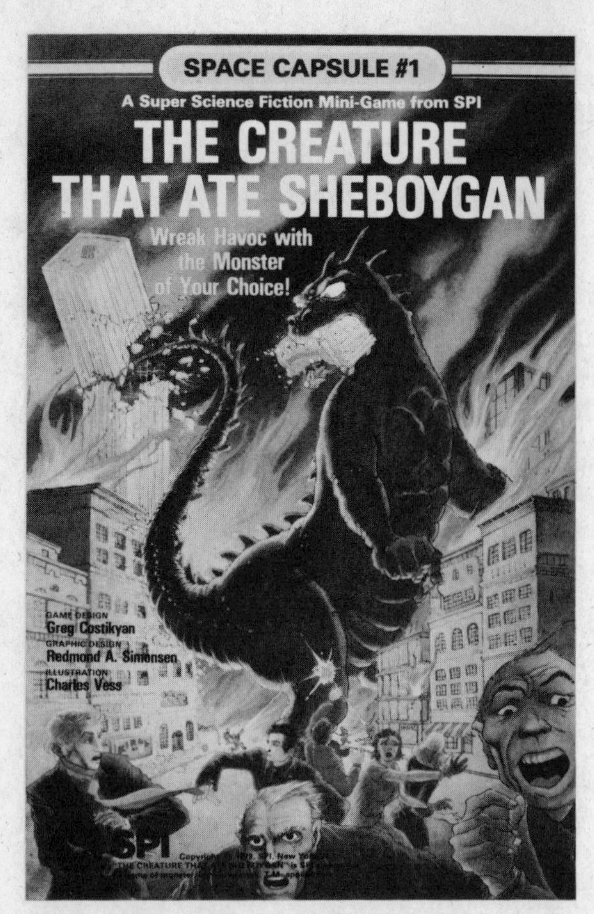

SPI's minigame The Creature That Ate Sheboygan. (PHOTO: SIMULATIONS PUBLICATIONS, INC.)

romaniacs will go wild. One advantage this game has over *Ogre* is that changing the combination of powers yields a significant difference in flavor and strategy. There are a lot of possibilities, and the special-powers list cries out for augmenting. If you don't take your wargaming too seriously, this is more fun than seeing the movie.

EVALUATION:
Presentation—Good
Rules—Fair to Good
Playability—Excellent
Complexity—5

OVERALL EVALUATION: Very Good

Dungeon! (1975)

PUBLISHER: TSR Games, Inc.

SUGGESTED RETAIL PRICE: $9.95 (boxed)

SUBJECT: Heroic fantasy adventure among various levels of a "dungeon."

PLAYING TIME: It takes two to more than five hours.

SCALE: This is a close-tactical game; each plastic piece represents a single individual adventurer. Each "square" represents something like five to ten feet.

SIZE: The 24" x 30" game uses "squares" of somewhat irregular size and shape.

BALANCE: The game favors Wizards and Superheroes over Heroes and, especially, Elves.

KEY FEATURES: Even more specifically than *King Arthur's Knights,* this is a board game introduction to fantasy role-playing games (and can, with minor adjustments, be refereed and played like one). Each of up to twelve (!) players takes the part of one of four sorts of adventurers; their common (but not joint) object is to be the first to return to the Main Staircase with the required amount of treasure. This amount varies depending on the power of the character. Monsters and treasure are distributed among the various rooms of the dungeon on separate cards. Each monster card has listed the numbers the different types of adventurers have to roll on two dice to kill it. If the monster survives the first attack, the adventurer must roll on the Combat Losing Table (CLT) to determine the adverse effects, which may involve the loss of a turn, an item of treasure, or the life of the hero. The six levels are graded, so that the nastiest monsters and the richest treasure are on the deepest levels. Wizards can also carry spell cards (to use instead of normal combat), and there are provisions for various magic items. Advanced rules allow players to ambush one another for the purpose of stealing treasure.

COMMENTS: This is a fine introduction to role-playing games, understandable by almost anyone, and helpful preparation for the far more complicated games of the next chapter. On its own, it has quite a few problems. Without the ambush rules, there is *no* player interaction at all; with them, player interaction takes over the game completely, and no one ever gets anywhere. The ambush rules are sufficiently vague—and poorly thought out—that an ambush inevitably becomes one long-drawn-out melee that is almost impossible to stop—at least if more than two players are involved (which is usually the case, sooner or later). Furthermore, despite the attempts to balance things by varying the treasure required, the more powerful characters can easily eliminate the Elves and Heroes from contention by taking a quick swing through the easier levels and clearing them of treasure. This is fun once or twice, but it gets old fast. Yielding to the temptation to tinker with the rules leads you straight to the bottomless pit of the next chapter.

EVALUATION:
Presentation—Very Good
Rules—Fair
Playability—Excellent
Complexity—3

OVERALL EVALUATION: Fair to Good

John Carter, Warlord of Mars (1979)

PUBLISHER: Simulations Publications, Inc.

SUGGESTED RETAIL PRICE: $15 (boxed)

SUBJECT: Heroic adventures on a fantasized Mars ("Barsoom"), adapted from novels by Edgar Rice Burroughs.

PLAYING TIME: The game takes three to five hours.

SCALE: This is basically a close-tactical simulation with a strategic movement display. Counters represent individual men and monsters. Time and distance scales are abstracted.

SIZE: The 22″ x 34″ game map employs both a strategic map of Barsoom (for long-distance movement) and a variety of tactical displays that use 3mm (⅛″) squares (each counter occupies a four-by-four array of such squares).

BALANCE: Good, but the outcome is often dependent on luck.

KEY FEATURES: Each player (there may be more than two) chooses from a selection of heroes and sets off in search of the foul villain who abducted his true love. The hero moves from city to city in an attempt to find the exact location of the villain and heroine. If he is discovered, he is usually attacked by evil minions, and the villain may flee to a safer city. The frequent melees can result in the hero's making friends with other warriors, who may then accompany him on his quest. Random events cards introduce treachery, miraculous escapes, and timely intervention by friendly forces, and there is a hilariously accurate Unspeakable Acts Table. The quests continue until all heroes in the game have overtaken and dealt with their respective villains, at which time each hero's overall performance is rated to determine the game winner. A relatively simply military game is also included, but it's of little interest compared to the rest of the package.

COMMENTS: This is even farther from a conventional wargame than most of those in this chapter; there is little interaction between players, and the luck factor is quite high. On the other hand, it's quite suited to solitaire play, and the flavor of the whole thing is very authentic Burroughs; the hero must overcome a seemingly endless series of adverse events only to have the villain slip away again and again. If the Burroughs style of improbable adventures pleases your palate, you'll eat this one up.

EVALUATION:
Presentation—Good to Very Good
Rules—Good
Playability—Very Good
Complexity—7

OVERALL EVALUATION: Good

King Arthur's Knights (1978)

PUBLISHER: The Chaosium

SUGGESTED RETAIL PRICE: $10 (resealable plastic bag)

SUBJECT: Chivalrous adventure in the romanticized Britain of King Arthur.

PLAYING TIME: Two to six hours are needed, depending on the scenario and the number of players involved.

SCALE: This would have to be termed a tactical-level game, but everything's pretty abstract, and movement is by areas rather than hexes.

SIZE: There is a 22″ x 34″ game map.

BALANCE: Good, although chance is a significant factor.

KEY FEATURES: Essentially, this is a race-and-chase game. Players each take the part of one of three levels of knights and compete to see who can first acquire treasure and Chivalry Points and return to Camelot. Encounters with nonplayer characters (and their treasures) are handled by means of various decks of cards. Knights may choose to wander around Britain, looking for a lucky card in any of the decks, or they may elect to draw a card from the Adventure deck and attempt to gain honor and points by fulfilling a quest: anything from righting a simple wrong to seeking the Holy Grail. It's possible to hinder an opponent by engaging him in personal combat, but this is neither a required nor a common occurrence. Characters (from Arthur and Guinevere to Merlin and Nimue), treasures, and events are all in keeping with the theme of legendary Britain.

COMMENTS: This is a direct descendant of a long-forgotten board game called *Prince Valiant,* and it has as much in common with *Chutes 'n' Ladders* as it does with *Tactics II.* It is, in effect, an introduction to role-playing games in a board game format. It's a lighthearted game, devoid of violence and horrible beasties, and quite suitable for children—if they have someone explain it to them. (It is *not* a childish game, only an idyllic one.) Unfortunately, the charts and tables are scattered hither and yon about the (rather long) rules, and except for the attractive four-color map, the components are rather primitive.

EVALUATION:
Presentation—Fair to Good
Rules—Good
Playability—Very Good
Complexity—5

OVERALL EVALUATION: Good

Nomad Gods (1977)

PUBLISHER: The Chaosium

SUGGESTED RETAIL PRICE: $10 (resealable plastic bag)

Cover for King Arthur's Knights game. Copyright 1978 by Greg Stafford. Game published by The Chaosium. (PHOTO: THE CHAOSIUM)

SUBJECT: Land combat on the Plains of Prax (on the imaginary world of Glorantha) between nomadic tribes.

PLAYING TIME: Two to five hours are required.

SCALE: Call this one an operational-level game; unit sizes and time and distance scales are not specified (and difficult to infer).

SIZE: The 22″ x 34″ game map uses 16mm (5/8″) hexes.

BALANCE: This is good.

KEY FEATURES: This is the second game in a series that began with *White Bear and Red Moon*, which was set on the same imaginary world (also the setting of next chapter's *Runequest*). Unlike the earlier game, this one is for two to four players, each of whom controls a strange and different band of nomads. The objectives are not territory but things: herds, personal possessions, magical artifacts. The Basic Game is just another battle, with little to distinguish it. The Advanced Game adds quite a bit of flavor and complexity, most of which involves keeping track of which units do what. To help things along, it's advisable to index all the counters with a page reference. While the rules and systems are not all that complex individually, there are so many of them that the effect is an undeserved complexity.

COMMENTS: The design and development of this game shows an understanding of fantasy and science fiction beyond the range of most other games. Most of the seventy-two pages of explanation are devoted to placing the units within the perspective of the rationale. The world hangs together nicely, but it's consistent, not self-explanatory. This makes it quite difficult for new players. The illustrations and unit markings are an aid to remembering what's what, but in play it's difficult to remember the difference between the Three Bean Circus and Tada's Sandals. If you are willing to spend the time and effort to become familiar with the world of Glorantha, you'll find *Nomad Gods* or *White Bear and Red Moon* interesting and enjoyable; most people, though, when faced with such a wealth of strange detail and an absence of familiar reference points, will find the going too tough.

EVALUATION:
Presentation—Good
Rules—Fair to Good
Playability—Fair to Good
Complexity—8

OVERALL EVALUATION: Good but highly specialized

Sorcerer (1975)

PUBLISHER: Simulations Publications, Inc.

SUGGESTED RETAIL PRICE: $14 (boxed—less in other packaging)

SUBJECT: Magical conflict on the imaginary world of Bannorkhemea.

PLAYING TIME: It takes three to six hours, depending on the scenario.

SCALE: This is an operational-level game. Counters represent individual sorcerers and legions of demons, trolls, and air dragons. Hexes equal ten kilometers, and each turn spans two days.

SIZE: The playing surface is 22″ x 34″ and hexes are 25mm (1″).

BALANCE: This is very good.

KEY FEATURES: There's a classic game in *Sorcerer:* it's called "Scissors-Paper-Stone." Onto that basis has been laid an involved (though not enormously complex) game of magical combat in another universe. The mapboard looks like Picasso's dropcloth; it's printed with hexagons in eight different colors, which are the game's key feature. Players control counters representing various magicians, each of whom controls one or more "colors" of magic, with which he can conjure units of dragons, trolls, and demons (essentially, cavalry, armor, and infantry) to fight the units of other sorcerers. In scissors-paper-stone fashion, colors of magic are arranged in a cyclical hierarchy. The relationship of colors determines the column of a differential Combat Results Table to decide battle results. Magicians may teleport—and conjured units may appear—only in properly colored hexes. Objectives vary from scenario to scenario but usually involve the acquisition of territory.

COMMENTS: This is a strange combination of the different and the ordinary. On the one hand, the basic premises—colored hexes, conjured units, teleportation—are different enough to be confusing. On the other hand, magical units all fight by conventional rules and resolve combat on a conventional CRT. Sorcerers fling magic bolts as if they were artillery. Sorcerers must expend movement points to accomplish tasks—a procedure similar to that of *StarSoldier* or even *Sniper!* The game is thus both hard to understand and hard to believe. Once the system is grasped, the presence of optimum strategies makes play stereotyped.

EVALUATION:
- Presentation—Very Good
- Rules—Good
- Playability—Good
- Complexity—7

OVERALL EVALUATION: Fair to Good

Swords & Sorcery (1978)

PUBLISHER: Simulations Publications, Inc.

SUGGESTED RETAIL PRICE: $15 (boxed)

SUBJECT: Fantastic adventure and conflict among various strange beings in an imaginary world (far, far away).

PLAYING TIME: Two to four hours are needed for the Quest Game; four to eight hours for the Army Game.

SCALE: The Quest Game could be termed tactical and the Army Game operational. Matters of scale, including unit size, are treated quite explicitly in the rules—but not in comprehensible English terms. For example, a turn equals one "zelith."

SIZE: The 22″ x 34″ game map uses 16mm (5/8″) hexes. There is a separate 8½″ x 22″ solar display.

BALANCE: This varies by scenario but is generally acceptable.

KEY FEATURES: As in *War of the Ring*, there is both a Quest Game and an Army Game, which may be combined—although the rules aren't too clear on how to go about it. As in other quest-type games of this chapter, the Quest Game is a race to retrieve certain objectives (assassination or rescue). In the Army Game, characters act as leaders, and combat is on a larger scale. There are many scenarios for either game, strange and varied races and monsters, and provisions for various arcane things from hidden movement to magical vortexes or reincarnation, including items recognizably borrowed from every other SPI game of this genre.

COMMENTS: The most important point about *Swords & Sorcery* is that it is not intended, even vaguely, to be a serious game. With characters like Unamit Ahezredit and Gygax Dragonlord (a reference to the head of TSR Games), terrain features like New Orc City and the Stream of Consciousness, and beings like killer penguins or the Corflu Cultists, it is hard to imagine anyone's missing this point, but many reviewers—and gamers—seem to have done so. This is as much a toy as it is a game. It is an enormous joke, and one you're invited to join in; the dozens of scenarios only begin to suggest the possibilities. Everything you can imagine—and pretty much everything SPI has ever tried in fantasy games—is here, and, unsurprisingly, things don't always fit together too well. The scenarios, for instance, often call for units that don't exist to set up in provinces that don't either. A bit of restraint all around would have helped make this a better and more coherent game and one more acceptable to the majority of gamers, but if you enjoy the designers' brand of humor, you'll have a lot of fun with this.

EVALUATION:
- Presentation—Excellent
- Rules—Good
- Playability—Good
- Complexity—8

OVERALL EVALUATION: Good

War of the Ring (1977)

PUBLISHER: Simulations Publications, Inc.

SUGGESTED RETAIL PRICE: $16 (plastic box; also available with the folio games *Sauron* and *Gondor* as *Middle Earth* for $22)

SUBJECT: The quest to destroy the One Ring and/or the attendant military battles, from Tolkien's *The Lord of the Rings*.

PLAYING TIME: It takes two to three hours for the Quest Game; four to six hours for the Army Game.

SCALE: The quest version is a tactical game, effectively; the army version is more a strategic-level game. Units are individual characters and (in the army version) armies. Turns represent the passage of one week.

SIZE: Three 11″ x 34″ game maps form a 33″ x 34″ playing surface. Hexes are 19mm (¾″).

BALANCE: The Fellowship player is favored, especially in the Quest Game.

KEY FEATURES: *War of the Ring* is two, two, two games in one. In the Quest Game, the familiar characters of the Fellowship play hide-and-seek with the Nazgûl and Orc search parties of Sauron, while collecting magic items with which to accomplish their ultimate goal: throwing the ring into the crack at Mount Doom. Semihidden movement with inverted counters for the Fellowship player aids in confusing the searches. Sauron is further limited by random search cards and a random (and very modest) task-point allotment with which he must move his units and conduct additional searches. Combat is handled very abstractly (though effectively) as one-on-one duels to the death. Many of the incidents of the book take place as a result of random events cards. In the army version, the characters act as the leaders of the respective armies, and rules dealing with mass combat, mobilization, and sieges are introduced, along with new events, various special provisions, and new victory conditions. There is even provision for a third player, Saruman.

COMMENTS: The richly detailed world of Middle Earth is an excellent setting for a wargame; it has a potential that no game designer could ever hope to realize fully. *War of the Ring* is a valiant effort at pro-

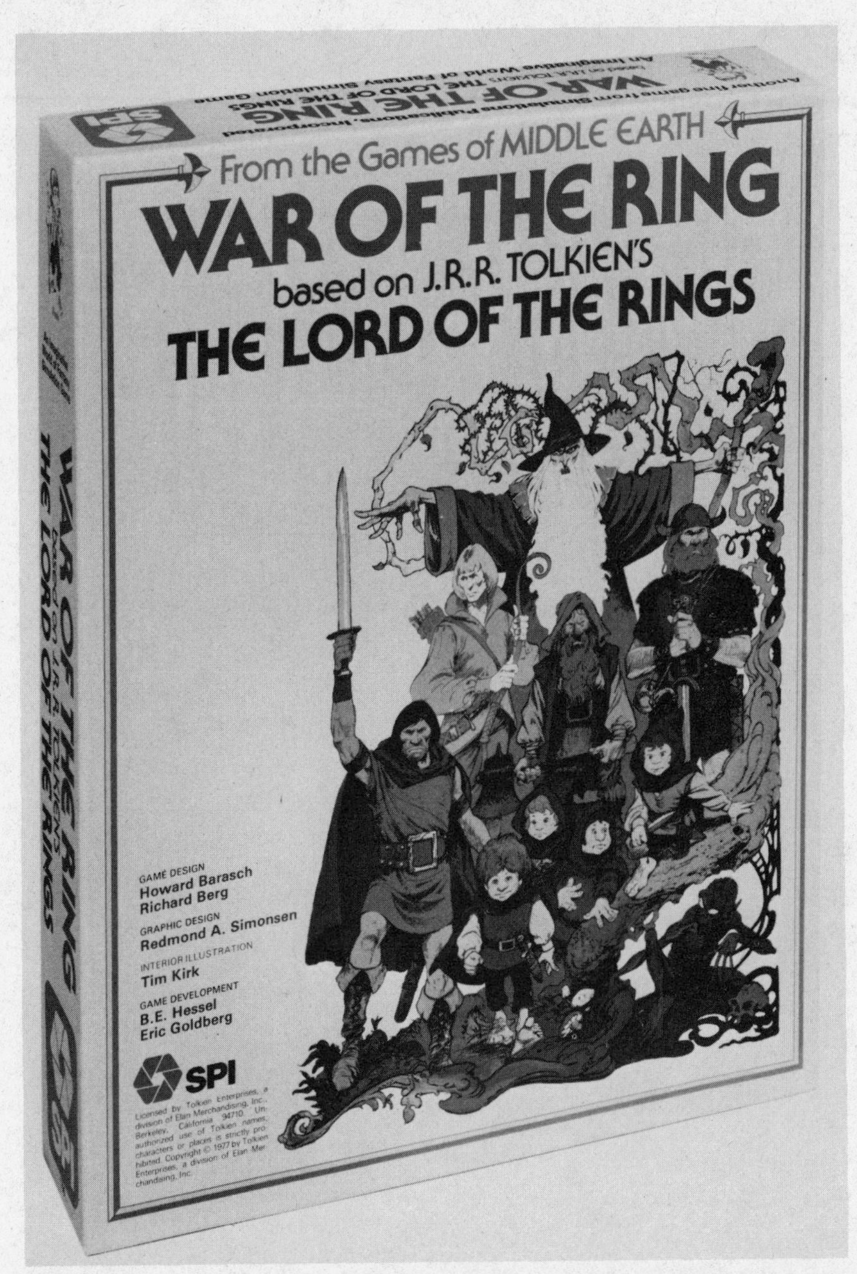

SPI's War of the Ring *game based on J. R. R. Tolkien's* The Lord of the Rings. (PHOTO: SIMULATIONS PUBLICATIONS, INC.)

ducing a workable simulation that retains the feel and color of Tolkien's setting. The game is aimed at introducing Tolkien fans to wargaming rather than the reverse, and there are enough holes in the rules that gamers unfamiliar with the plot may have some difficulties. The simpler Quest Game is more flexible—and thus more successful—than the longer Army Game. This is a decent game in its own right and a must for Tolkien addicts. Tim Kirk's artwork on the map and character cards is marvelous.

EVALUATION:
- Presentation—Very Good
- Rules—Fair
- Playability—Good
- Complexity—Quest Game: 5; Army Game: 7

OVERALL EVALUATION: Very Good

CHAPTER 16

Dungeons, Dragons, and Role-Playing Games

PARADOXICALLY, ROLE-PLAYING games are both the oldest and the youngest members of the wargame family. They are a revival of miniatures, which were overwhelmed in the United States with the advent of board wargaming, and of *kriegspiel's* "nonplaying referee," who had disappeared even from military wargaming after World War II. Yet they are also something quite new, and didn't exist in commercial form before 1974.

The phenomenal growth of role-playing games (RPGs) makes conventional wargaming's quite rapid expansion look like the slow accretion of a coral reef. Led by the now famous *Dungeons & Dragons,* published by TSR Games, Inc., science-fiction and fantasy role-playing games have come out of nowhere to the point that, in five short years, there are now far more miniatures devoted to them than to all other wargaming pursuits combined. Virtually singlehandedly, *Dungeons & Dragons* established and continues to sustain, generally, what is essentially an entire new hobby and, specifically, what is the third largest company in the business.

With few exceptions, a role-playing game cannot be opened up, learned, and played in the normal way. It doesn't even look like a conventional game: there's no board, no counters, and, usually, no dice, although they are required for play. There's just a rule book or three. Even in theory, a "game" like *Dungeons & Dragons* is less a game than a game system, and in practice it's less *that* than a system for designing a game system! In a general sense, it's an approach to treating the fantasy subjects of the previous chapter on a man-to-man level. Straight historical settings have been tried occasionally, as in *En Garde!,* but such games apparently do not have enough of the imaginative—the unexpected—to suit RPG fans.

Basically, a role-playing game system involves two parts: a character-generation system and a conflict-resolution system. Everything else is trimming—though there's usually plenty of that, and it's trimming, often in the setting, that helps separate one RPG from another. In *Dungeons & Dragons* this trimming includes charts and tables of magical and clerical "spells," monsters that the player may run into (usually "adapted"—to put it most charitably—from myth, folklore, or fiction, but occasionally invented, generally without regard to logic or biology), traps (everything from pits to poisoned needles), and treasure. The chief differences between science-fiction and fantasy role-playing games are in the nominal settings (a pseudomedieval land of knights and dragons versus the interior of a large starship, for instance) and in the flavor of the condiments: a manticora versus a Martian land shark, a magic carpet versus an antigravity belt, a sleep spell versus a sleep grenade.

Since *Dungeons & Dragons* was the first successful commercial role-playing game (private play-by-mail campaigns date back a good fifteen years), and since more people play what are, at least nominally, *Dungeons & Dragons* compaigns than all other RPGs put together, our subsequent comments will tend to focus on *Dungeons & Dragons.* Except in matters of detail, most comments will also pertain to other such games.

While the magnitude of their success is difficult to understand fully, even in retrospect, role-playing games clearly came about because of a need to handle personality and imagination—the two most important elements of fantasy, which are very difficult to treat in board games. Personality was supplied through the use of incredible detail and a single playing piece (a "character") per person, which permitted complete identification between a player and his game alter ego. Imagination came in through the use of entirely open-ended rules and a game system that required the referee to develop his own setting, his own characters—his own world.

Any number of people can play a role-playing game, although a dozen at a time is about the practical limit even for an experienced referee—usually called a dungeon master or DM. Normally, each player controls—and takes the part of—a single character, a "person" whose specific capabilities are a product of six attributes (in *Dungeons & Dragons;* names and numbers vary in other systems): Strength, Constitution, Dexterity, Intelligence, Wisdom, and Charisma. Most of these are fairly self-descriptive—and the ones that aren't are not satisfactorily described (or used) in the rules, anyway. These attributes are determined by rolling three dice for each; this gives a nicely bell-curved range of 3 (low) to 18 (high). After surveying these, the player decides whether he wishes his character to be an elf, dwarf, hobbit—excuse us, "halfling"—magic-user (a typically graceless term for one who uses magic: a magician or mage), cleric (priest), fighting man, or thief. These various types are supposed to be archetypes of the characters found throughout heroic fantasy: a fighting man would be a sword-wielding hero like Conan or Aragorn; a cleric would be of the Friar Tuck sort; a magic-user would resemble Merlin or Gandalf and would, of course, specialize in the casting of magic spells. A fighting man needs good Strength and Constitution; a magic-user should have high Intelligence.

After a vocation/race and a name (anything from Beowulf, John Carter, or Fearless Fosdick to Cardamum, Gardol, or Quasar) are chosen, the character gets a certain number of "hit points." This central element, which is *not explained* in the rules, is a bit like a combat factor and represents specifically the amount of damage (in points) the character can withstand before dying. This, inexplicably enough, has almost nothing to do with Constitution (literally nothing by the standard, unsupplemented rules); instead, like practically everything else in *Dungeons & Dragons,* this comes from a separate die roll. The *size* or type of die depends on the vocation of the character. For example, a cleric is allowed a standard six-sided die; a fighting man rolls an eight-sided die; a feeble old magic-user (who may have a Constitution of 18!) is limited to a four-sided die. These dice, which are essential to the play of most role-playing games, are not usually included with the game. Depending on the policy of the DM, the hit points may or may not be known to the player; since they are not explained and barely mentioned, there's plenty of room for interpretation. (This could be taken to be a capsule description of almost anything in *Dungeons & Dragons*.)

After obtaining whatever other intangible impedimenta local custom requires (height, weight, birthday, age, god, etc.), each player must spend his money (given by another die roll) to equip his character with weapons, armor, food, clothes, horses, packs, and so on. Imagine yourself setting out into unexplored terrain known to be inhabited by ghouls, vampires, dragons, werewolves, giant thises, flying thats, and poisonous the others—and then imagine what you might carry: it's probably available.

After all this (which, as you may have guessed, may take hours the first time), you are finally ready to move.

Dungeons & Dragons was directly derived from miniatures, which makes it confusing for people familiar only with board wargames. Hexes are not the basic unit of measurement; instead, all distances are given in inches—which translate, somewhat confusingly, into *tens* of feet (underground or inside buildings) or yards (outside). Nor, as we mentioned earlier, is there a board in the conventional sense. Instead, a variation of the double-display system found in *Jutland* or *Bismarck* is used. The "strategic" map may itself be twofold: an overall map of the country or region (if it is to be a full-fledged campaign, complete with outdoor adventures) and various detailed drawings of each individual building—or each floor (a "level" in *Dungeons & Dragons* parlance) of each building—of any consequence. Any or all of these things may be termed "dungeons." These must all be created by the dungeon master. They are never shown to the players, who must make (not "make *up*") their own map during the course of play, usually on graph (inside) or hex (outside) paper.

Everything seen, heard, or sensed by the players is relayed to them by the DM. Appropriate topography—from corridors and rooms of such-and-such a length to swamp or forest as far as the eye can see—is then transferred onto their map, which will then help guide them on the way back and during future sessions.

When a "tactical" display is needed—when the adventurers encounter other characters or monsters with whom they have to interact physically (meaning, generally, in combat)—miniature figures, which must be obtained separately, are used. Each represents an individual character or monster; together, they are maneuvered on any convenient flat surface. As a visual aid, graph paper, hex paper, or a checkered tablecloth may be used; otherwise, a ruler is required to move everyone properly.

In role-playing games, unlike any other wargames and, indeed, unlike virtually all other games, *period*, the players don't normally compete with or battle each other. Instead, they cooperate to varying degrees; they play "against" fate, chance, nature, and an arkload of nonplayer characters (NPCs: RPGs are full of initials) and monsters, all of which are controlled by the DM, who is thus

Cover from Sorcerer's Apprentice, *a quarterly magazine for fantasy-gaming fans that is edited by the designer of* Tunnels & Trolls. *© Flying Buffalo.*

designer, board, referee, and opponent, all at once!

When the group is brought into conflict with a dragon, say, the battle is resolved something like a tactical ancient or medieval board game. Anybody who is in range can cast a spell, fire an arrow, swing a sword, throw a dagger, or make faces at the dragon; each attack is resolved separately to obtain the number of points of damage done to the dragon. At the same time, the dragon is clawing, biting, or breathing fire on one or more of his assailants, who take points of damage in return. Generally, this continues until the dragon or everyone in the group is dead.

The key and interesting point here is not just that there is a combat system that handles this (or a magic system to handle *that*); it's that there is a general system of conflict resolution. With the application of a little imagination, it can handle *anything*. In effect, any action that is possible (given the nature of the fantasy world and the particular abilities of the individual character) is given a chance of happening (according to a chart or table or an arbitrary determination by the DM), and a die is rolled to determine the result. To succeed in hitting the dragon with an arrow, a bowman might be required to roll a 15 or higher on a twenty-sided die; if he does, he rolls another die to determine how much damage he did to the dragon. Conversely, if the dragon breathes fire at the group, a potential victim is allowed to make a "saving throw" to see if the potential calamity fails to affect him; if the character "fails" his saving throw, the magical/extraordinary attack/disaster succeeds, and the character is in Big Trouble.

The players are not, thus, limited in their actions; nor does one player's actions determine another's. While the brave-but-not-too-bright hero is poking the dragon with his spear, a less reckless fellow might be trying to climb a tree, hide in the shadows, or negotiate a peaceful settlement. Using guidelines given in the rules or developed through experience, the DM can, by rolling appropriate dice, decide how well and how soon all of these actions take place.

Survivors of encounters with dragons or land sharks get whatever treasure the beasts may have been guarding—probably enough loot to buy fancier equipment for their next expedition and possibly some of the much desired magical items: wands, potions, crystal balls, or weapons. It's important to note that the characters get their share of the hoard "for keeps"—or until they spend it. But it goes with them the next time they play, along with everything else they ever acquire.

Dungeons & Dragons is not normally played just as an isolated game; each session is part of an ongoing campaign with no end. There is no "winner" and no criteria for winning. You just keep playing as long as you find it enjoyable—for years, if you're an RPG addict. Admittedly, accumulating riches is something of an aim for most players, but it's neither the sole nor the most important reason for playing.

More significant, even for scorekeepers, are "experience points," which are as poorly explained as anything else. This is a measure of the intangible benefits derived from experiencing adventures. These points, divided into "levels" (no relation to other uses of the term in the game), bear directly on some fairly significant factors, including both hit points (you get one die per level; a "seventh-level" character gets seven dice worth of hit points) and the ability to perform typical actions. Fighters become more proficient with their weapons; mages learn more and more powerful spells; thieves get better at picking locks, and so forth. In keeping with the open-endedness of the game, there is no limit to how high a character may progress; however, a willing suspension of disbelief becomes progressively more difficult. The system itself shows signs of strain when anything higher than 10th- or 15th-level characters are involved, and total absurdity sets in not long after. Despite the well deserved success of the movie, Superman has always been an unsatisfactory character because he is too powerful; a fighting man who can survive ground-zero fireballs or dispatch a dragon with his pinky is, from a human perspective, meaningless. (Most of the chief alternatives to *Dungeons & Dragons* share the same weakness.)

Although the mechanics of role-playing games—even more than conventional board wargames—are based on the possibility of quantifying everything from combat to temperamental reactions, playing in a mechanical way is akin to a two-player game of *Stellar Conquest* (or kissing your sister, to drag in an appropriately hoary cliché): it's better than a broken leg, but you're missing half the fun. Maximum enjoyment for all concerned comes only from imbuing your character with a fully developed personality appropriate to his attributes. Many marvelously imaginative campaigns have dried out, and sometimes died out, because of a lack of this vital element.

There is a seemingly endless flow of material from various companies supplementing any basic set of charts and lists: not only new selections of monsters and treasure but everything from blank, modular dungeon sections to entire cities—detailed, populated, and coded for random encounters! In the face of such variety and a natural desire to innovate, it is commonplace for a dungeon master to borrow bits and pieces from this novel, that supplement, the other game, and his own imagination; when integrated into a campaign, the unique result may only distantly resemble any one set of rules. Fortunately, uniformity is the one element that is neither required nor desired in a role-playing game.

In the end, the success of a role-playing game is least dependent on the particular set of rules that it's based on. What's more important is the attitude of the players and their willingness to join in the spirit of adventure required. Most important of all, however, is the dungeon master, on whose personality, imagination, and judgment everything depends. A good DM can use the poorest set of rules to create a delightful adventure, while in the hands of an inept referee the best game will be doomed to mediocrity.

A WORD OF WARNING

Although an attempt has been made to evaluate role-playing games on the same scale as the others covered in this book, it's dangerous to take cross-comparisons too literally. This is particularly the case with the Complexity and Playability ratings. Once an RPG campaign is in progress, it's a reasonably simple matter for a new player to join. Since much of any RPG system is invisible to the players, the newcomer may have to learn only a few of the rules and none of the charts and tables. For the aspiring dungeon master, however, beginning a real campaign is an undertaking that makes *Drang Nach Osten!* or *Campaign for North Africa* look simple.

For much the same reason, without prior experience you cannot play—or even understand—most role-playing games just from reading a set of rules, which are often poorly written and full of holes King Kong could waddle through. With few exceptions, you must find an existing campaign and be introduced to play by other players. Happily, this is no longer difficult. After gaining some practical experience and an understanding of how a campaign is run, it is relatively easy to tackle any of the role-playing games in print.

Evaluations

Boot Hill (1975)

PUBLISHER: TSR Games, Inc.

SUGGESTED RETAIL PRICE: $9.95 (boxed)

SUBJECT: Role-playing rules for battles and/or a campaign set in the Wild West of the late 1800s.

KEY FEATURES: Four characteristics—Speed, Bravery, Accuracy, and Strength—are determined by rolling percentile dice giving numbers 1–100. Speed—and, to a lesser extent, Bravery—determines the precise order in which combat is resolved. Accuracy is important, since a hit-location system is used to determine the severity of the wound. Strength, the equivalent of Constitution in some other games, represents the ability of the character to withstand damage. Since all combat results take effect immediately, the order of fire is extremely important—and quite time-consuming to calculate, as various modifications are involved. The Advanced Game introduces written plots for simultaneous movement and, if a referee is present, hidden movement. Other sections introduce nonplayer characters. Optional rules, which add greater realism at a considerable cost in playability, include an order of combat that integrates intermittent fire, various subclassifications of what is already greased lightning, aimed fire, gamblers, intoxication, stray bullets and misfires, exotic weapons, a brawling system for less-bloodthirsty characters, posses, and tracking. Two sample scenarios are provided: the gunfight at the OK Corral (of course) and the simultaneous robbery of two banks in Coffeyville by the Dalton gang. Other suggestions for scenarios include bounty hunting, Indian fighting, stagecoach robbing, and many others.

COMMENTS: *Boot Hill* is a game that is well suited to portraying small battles based on the Old West. Unfortunately, the mechanics are such that the playing time required increases exponentially with the number of individual figures involved—especially if the alternate first-shot determination is used. The brawl system, which tends to produce knockdown dragouts more reminiscent of the theater than of real life, is seldom a consideration unless ammunition runs low. Somewhat more consideration is given a campaign here than in, say, *Space Patrol*—but not much more. There is certainly no scarcity of reference materials, however, and the system can be adapted to suit virtually any literary western setting and any heroic character. A real campaign would severely test the judgment and resources of the referee and the spirit of the players, but, because the milieu is so familiar, "dropping in" for an occasional battle/scenario is a possibility.

EVALUATION:
Presentation—Good
Rules—Good
Playability—Fair to Good
Realism—Very Good
Complexity—8

OVERALL EVALUATION: Good

Chivalry & Sorcery (1977)

PUBLISHER: Fantasy Games Unlimited

SUGGESTED RETAIL PRICE: $10 (one *large* rule book)

SUBJECT: Role-playing rules are provided for a large, detailed campaign in a medieval setting, with or without the element of fantasy.

PLAYING TIME: This is discretionary but very long.

KEY FEATURES: Over five hundred pages of rules have been compressed and reduced into 128 pages of fine print guaranteed to drive you blind or mad. It contains, obviously, an enormous amount of detailed information for the establishment of medieval campaigns, including mass actions, sieges, and full-scale wars as well as jousts or dungeon adventures. The elaborate combat system offers many options and includes modifications for everything but tooth decay. The multivarious magic systems are mostly in keeping with medieval or ancient precedent. Characters get nine primary attributes (rolled with twenty-sided dice)—*Dungeons & Dragons'* six plus Personal Appearance, Bardic Voice, and Alignment—and a flock of sometimes highly involved and/or important characteristics like Social Level.

COMMENTS: In early, privately run play-by-mail RPGs, players typically took the part of great lords, princes, or counselors who controlled the affairs of a kingdom. *Chivalry & Sorcery* is aimed at just that kind of grand campaign. Played by mail at a suitably leisurely pace, with the designer as referee, this might possibly be struggled through; it certainly isn't playable in the ordinary sense of the word. Even if the question of complexity were disregarded, it's not suited for dungeon adventures of the conventional sort, and indeed the monsters and most of the fantasy material are included almost as an afterthought. The magic systems are interesting, well researched, and fairly "realistic," but they're wildly unsuited to normal adventuring. Kings and serfs may both have a place in an educational grand campaign—but not on the same adventure. Using a twenty-sided die to generate characteristics gives much too random a result, but that's relevant only if you were planning to play this—which is obviously absurd. However, as a source of material to incorporate into a saner and more exciting RPG, *Chivalry & Sorcery* compares favorably with the *Encyclopaedia Britannica*.

EVALUATION:
Presentation—Good
Rules—Very Good
Playability—Very Poor
Realism—Excellent
Complexity—10

OVERALL EVALUATION: Poor as a game; Excellent as a source

Dungeons & Dragons (1974)
PUBLISHER: TSR Games, Inc.
SUGGESTED RETAIL PRICE: $14.95 (three booklets, boxed)
SUBJECT: Fantasy role-playing rules are provided for a campaign set in a pseudomedieval milieu heavily influenced by J. R. R. Tolkien's *The Lord of the Rings*.
PLAYING TIME: This is entirely at the discretion of the referee and players; typically, four to eight hours at a sitting.
KEY FEATURES: This is *the* standard system. It derives in part from the fantasy section of TSR's otherwise uninteresting *Chainmail* miniatures rules. Players take the part of characters who are given attributes in six primary categories according to the rolls of three dice and separated into classes (vocations) that limit or define various of their abilities. Characters advance within classes from one "level" to another by accruing "experience points." Combat involves twenty-sided dice; the required result is based on the attacker's class and level (actually, group of levels) and the defender's armor class (a measure of armor type and the presence or absence of a shield). There is a separate damage roll after a successful attack. Sustainable damage is based on "hit points," which are dependent primarily on level and secondarily on class. The magic system, which is taken from Jack Vance's novel *The Dying Earth*, causes a magic-user to forget a spell as soon as it is cast. There are charts and descriptions of magical and clerical spells, monsters, and magical treasures. The various supplements expand and clarify much of this, particularly the essential *Greyhawk*, which adds an important character class (thieves), certain attribute-related effects, and many details.
COMMENTS: What can be said about a phenomenon? Aside from *Tactics II* and possibly *PanzerBlitz* (the first modern tactical wargame), this is the most significant war game since H. G. Wells. On the other hand, beginning characters are without exception dull, virtually powerless, and so fragile they would fare poorly against a crippled mosquito—none of which exactly encourages newcomers. To compensate for the inherent flaw in the stupid magic system, many of the spells are redundant, and the effects of the majority are hopelessly vague. Armor affects the likelihood of being hit rather than net damage sustained—a typically *anti*realistic absurdity. None of the numbers, such as those required to hit another character, relate logically to any others, such as the defending man's armor class; armor class numbers run *backward*—but not always (those of magic armor go backward *and* forward). The essential elements—saving throws, hit points, experience points, and so on—are undefined or poorly explained; the ratio of substance to "holes" compares unfavorably with the head of a tennis racquet. Presented in the most illiterate display of poor grammar, misspellings, and typographical errors in all of professional wargaming, the rules make *Fall of Rome's* look cogent. Some errors like "% liar" for the equally cryptic "% lair" have spawned the most bizarre misinterpretations. This and the generally poor systems have caused actual cam-

paigns to diverge widely from what is specified or suggested in the rules. Generally, this is for the better. As it was given birth, it is fascinating but misshapen; it its best incarnations, it's perhaps the most exciting and attractive specimen alive.

EVALUATION:
Presentation—Poor
Rules—Rock Bottom *
Playability—Fair
Realism—Fair
Complexity—7; with supplements, 8

OVERALL EVALUATION: As is, the "fountainhead"; as played, the best game going

Dungeons & Dragons, Basic Set (1977)

PUBLISHER: TSR Games, Inc.

SUGGESTED RETAIL PRICE: $9.95 (boxed)

SUBJECT: Fantasy role-playing rules are provided for an initial campaign.

PLAYING TIME: This is discretionary; typically, four to six hours.

KEY FEATURES: With the addition of a few new spells, this is straight *Dungeons & Dragons*—but only part of it. Essentially, this is a condensation and reorganization of material from the first three rule books plus the supplements that directly pertain to starting a *Dungeons & Dragons* campaign. Only things appropriate to the first three character levels are included, so all charts and tables are truncated. Also included are a set of play aids useful for the beginner: polyhedral dice, a monster and treasure list for random encounters, and a set of "dungeon geomorphs"—uncoded modular sections that can be arranged in different ways to construct a dungeon setting quickly.

COMMENTS: *Basic Dungeons & Dragons* is only a starter set and effectively obsolete a few weeks after you get a campaign going. It was written ("edited," if you prefer) by someone outside the TSR establishment who knew a noun from a verb, and the difference shows. It's still a long jump short of perfection, but you *can* read this and generally understand what's going on. The playing aids are useful, if only as examples. It's still preferable to participate in an ongoing campaign, but if you must venture into RPG country without a guide, this is the first place to visit.

EVALUATION:
Presentation—Very Good
Rules—Good
Playability—Good
Realism—Fair
Complexity—6

OVERALL EVALUATION: Good but incomplete

Dungeons & Dragons, Advanced (1978)

PUBLISHER: TSR Games, Inc.

SUGGESTED RETAIL PRICE: $9.95 (for each of four large-format hardcover books)

SUBJECT: Fantasy role-playing rules are provided for a complete campaign.

PLAYING TIME: This is discretionary; typically, five to eight hours.

KEY FEATURES: This is a more complicated and somewhat altered version of standard *Dungeons & Dragons* as augmented by the four supplements, articles in TSR's magazine *The Dragon*, and other material. There are more character classes, and their hit dice are a bit different. Attribute effects are more varied and important.

* Once again, we're forced to extend our range.

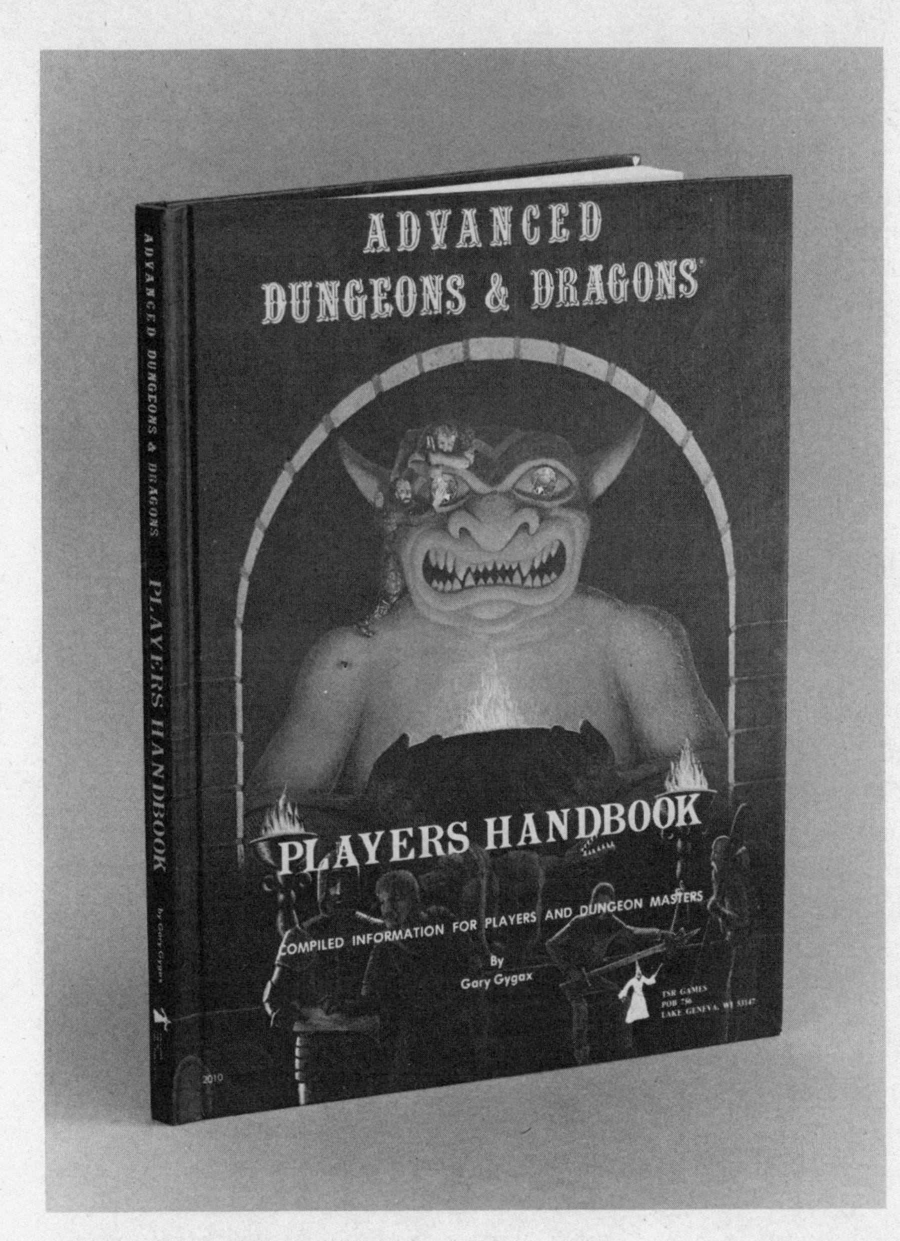

Players Handbook for Advanced Dungeons & Dragons. (PHOTO: TSR GAMES, INC.)

Armor classes have been reshuffled slightly (though they still run backward), and combat is *Greyhawk* with minor modifications. There are far more spells, most of which have verbal, somatic (for example, a gesture), and material components. The material has been organized into four books: the *Monster Manual, Players Handbook,* and the *Referee's Guide* have been issued; a rewritten version of *Gods, Demi-gods & Heroes,* is to follow in 1980.

COMMENTS: The biggest differences between this and the old *Dungeons & Dragons* are cosmetic: The books are handsome volumes; the typography is up to civilized standards; the language approximates standard English; as in basic *Dungeons & Dragons,* some of the terminology has been changed by request of the Tolkien estate ("hobbits" became "halflings," and "ents" turned to "treants," though by a bit of linguistic legerdemain "orcs" remain "orcs"). Changes in substance are much fewer: gaps have been filled, details added, and some use made of all of a character's six attributes. It is now possible to understand what the designers had in mind. This is unquestionably a richer, more satisfactory set of rules than the first. However, all of the inconsistency, illogicality, and awkwardness of the original flawed systems remain intact. It's hard to imagine publishing *two* new versions without admitting there was room for improvement in the old one, but that's what TSR has done. Their defense of their continued use of four different meanings of the term "level" is a marvelous bit of sophistry that adds up to "we did it that way the first time"—not much of an excuse in the face of changes in spells, character classes, hit dice, armor classes, attributes, terms, and other individually minor details that add up to something explicitly incompatible with existing *Dungeons & Dragons* campaigns. *Dungeons & Dragons* needed a complete overhaul; unfortunately, what it got was a minor tuneup and a new paint job.

EVALUATION:
Presentation—Very Good
Rules—Very Good
Playability—Fair
Realism—Good
Complexity—9

OVERALL EVALUATION: Good

Empire of the Petal Throne (1975)

PUBLISHER: TSR Games, Inc.

SUGGESTED RETAIL PRICE: $27.50 (boxed)

SUBJECT: Fantasy role-playing rules are provided for a campaign set on the imaginary world of Tekumel.

PLAYING TIME: As with most RPGs, a campaign can last forever; each session is typically four to eight hours.

SIZE: Two 22″ x 33″ maps and one 22″ x 28″ map all use 16mm (5/8″) hexes, but none are required for play.

KEY FEATURES: *Empire of the Petal Throne* presents a complete fantasy role-playing milieu that includes fighters, priests, magic-users, monsters, treasures, magical items, spells, a language, and a culture. The system is vanilla *Dungeons & Dragons* (it even uses the same combat table) with some of the terms and fauna changed to fit the very different setting. Two of the maps are a small-scale representation of the main continent of Tekumel; the remaining map shows the city in which new players start.

COMMENTS: This game is incredibly detailed, well thought out, and self-consistent. Although it uses the same basic game system as *Dungeons & Dragons*, the framework is better presented and put together. The culture upon which *Empire of the Petal Throne* draws is a creation of the fertile mind of M. A. R. Barker and, as such, is alien and unfamiliar to everyone else who comes in contact with it. That is both a strength and a weakness: an adventurer can experience something more novel and bizarre than is usually the case in the somewhat predictable fantasy version of the Middle Ages, but there is probably no one other than Barker who can adequately run a campaign. Most people who have the game have probably used it as a source of ideas for incorporation into their own more conventional fantasy worlds.

EVALUATION:
Presentation—Very Good

Rules—Good
Playability—Fair
Realism—Good
Complexity—8

OVERALL EVALUATION: Fair game but overpriced

En Garde! (1975)

PUBLISHER: Game Designers' Workshop

SUGGESTED RETAIL PRICE: $4 (booklet)

SUBJECT: Role-playing rules are provided for a campaign set in historical (if romanticized) seventeenth- to eighteenth-century Europe.

PLAYING TIME: This is discretionary but sessions are somewhat shorter than those of most RPGs; typically, two to four hours.

KEY FEATURES: Characters have four abilities: Strength, Constitution, Expertise, and Endurance—the first three die-rolled in standard fashion, and the last one literally a product of the first two. The fencing system—the only realistic one in RPGs—consists of a sequence of actions arranged in a series of set routines; damage, a product of Strength and a result drawn from an Attack/Defense mode matrix, is applied against Endurance. Instead of "experience points," a simpler system of "status points" is used. These are earned not only from duels but also from social activities like gambling, carousing, and pursuing a mistress. Like the combat procedures, social activities are planned (by the month, in weekly segments), plotted, and revealed simultaneously (as in *Diplomacy*). The general object is to accumulate enough status points each month to increase one's status level, which gets harder as you go along.

COMMENTS: No game that allows toadying as a social activity and recommends "death by stoning" for players (*not* characters!) caught cheating takes itself too seriously—a pleasant change from the heavy-handed pronouncements periodically issued by some competing companies. *En Garde!* began life as an attempt to create a realistic fencing system and became a role-playing game when the background, predictably, engendered more interest than the fighting. The social situation will not attract rabid women's liberationists, and the game's limited scope (no monsters, no magic, no hoards of treasure) will keep it from challenging *Dungeons & Dragons*, but it makes a delightful change of pace. The simple system is easy to referee, and except for *Melee* and *Wizard*, it's the only RPG you could imagine running concurrently with your major campaign, on, say, alternate Tuesdays.

EVALUATION:
Presentation—Good
Rules—Very Good
Playability—Very Good
Realism—Good
Complexity—5

OVERALL EVALUATION: Good

Gamma World (1978)

PUBLISHER: TSR Games, Inc.

SUGGESTED RETAIL PRICE: $9.95 (boxed)

SUBJECT: Role-playing rules are provided for a campaign set in North America in the twenty-fifth century, 150 years after a nuclear holocaust.

PLAYING TIME: This is discretionary; typically, three to eight hours at a sitting.

SIZE: The 16″ x 25″ map uses 6mm (¼″) hexes but is not required for play.

KEY FEATURES: The setting for this game is

Gamma World, *published by TSR Games.* (PHOTO: TSR GAMES, INC.)

a post–nuclear holocaust America, as rampant with radiation as *Metamorphosis Alpha's* starship, and indeed the vast majority of the systems are adapted from *Metamorphosis Alpha* (and, more indirectly, from *Dungeons & Dragons* as well). Combat has been elaborated by the inclusion of initiative, fatigue, and a greater variety of weapons and armor, and the mental and miscellaneous attack matrices have been cleaned up. There are more physical and mental mutations, but they're also better organized. Three unique flow-chart mazes have been devised to determine a character's success at figuring out newly found artifacts. Experience points are back, and the implicit levels can yield improvements in attributes or fighting abilities (sort of a cross between the systems used in *Melee* and *Runequest*). Also included are a campaign map of an altered North America and a set of polyhedral dice.

COMMENTS: This is in every sense an improved version of *Metamorphosis Alpha.* Rules have been tightened and elaborated, and the organization is clearly superior. All the charts, tables, and lists for random monster and treasure generation are included on tear-out sheets at the back of the book. The artifact mazes are a useful device, nicely done. A more detailed background includes thirteen secret societies with obvious possibilities for player interaction and campaign conflicts. While it's not quite as open-ended in scope as *Dungeons & Dragons,* it has plenty of potential. It's hardly flawless, but *Gamma*

World is one of the best things TSR has produced.

EVALUATION:
- Presentation—Very Good
- Rules—Good
- Playability—Good
- Realism—Good
- Complexity—7

OVERALL EVALUATION: Very Good

Melee (1977)

PUBLISHER: Metagaming

SUGGESTED RETAIL PRICE: $2.95 (plastic envelope)

SUBJECT: The game focuses on personal, man-to-man (and -monster) gladiatorial combat.

PLAYING TIME: It takes about thirty minutes to play.

SCALE: This is a game of close-tactical combat; each counter is an individual man, animal, or monster. Hexes are about six feet across and each turn represents a few seconds.

SIZE: The 8″ x 12″ "board" uses 25mm (1″) hexes.

BALANCE: The balance is excellent.

KEY FEATURES: This is strictly a combat system for individual battles or very limited adventures. Only two attributes—*not* randomly generated—are used: Strength (ST) and Dexterity (DX). Strength determines the weapons a character can wield and the amount of damage he can sustain. Dexterity determines the character's ability to hit an opponent; it is temporarily adjusted for movement, encumbrance, and the like. Weapons and armor effects are similar to those in *Tunnels & Trolls*, though combat resolution is quite different. All die rolls require six-sided dice only. The system in *Melee* and *Wizard* is the only one based on board games instead of miniatures rules.

COMMENTS: In *Melee*, designer Steve Jack-

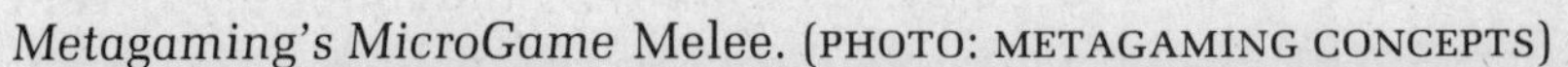
Metagaming's MicroGame Melee. (PHOTO: METAGAMING CONCEPTS)

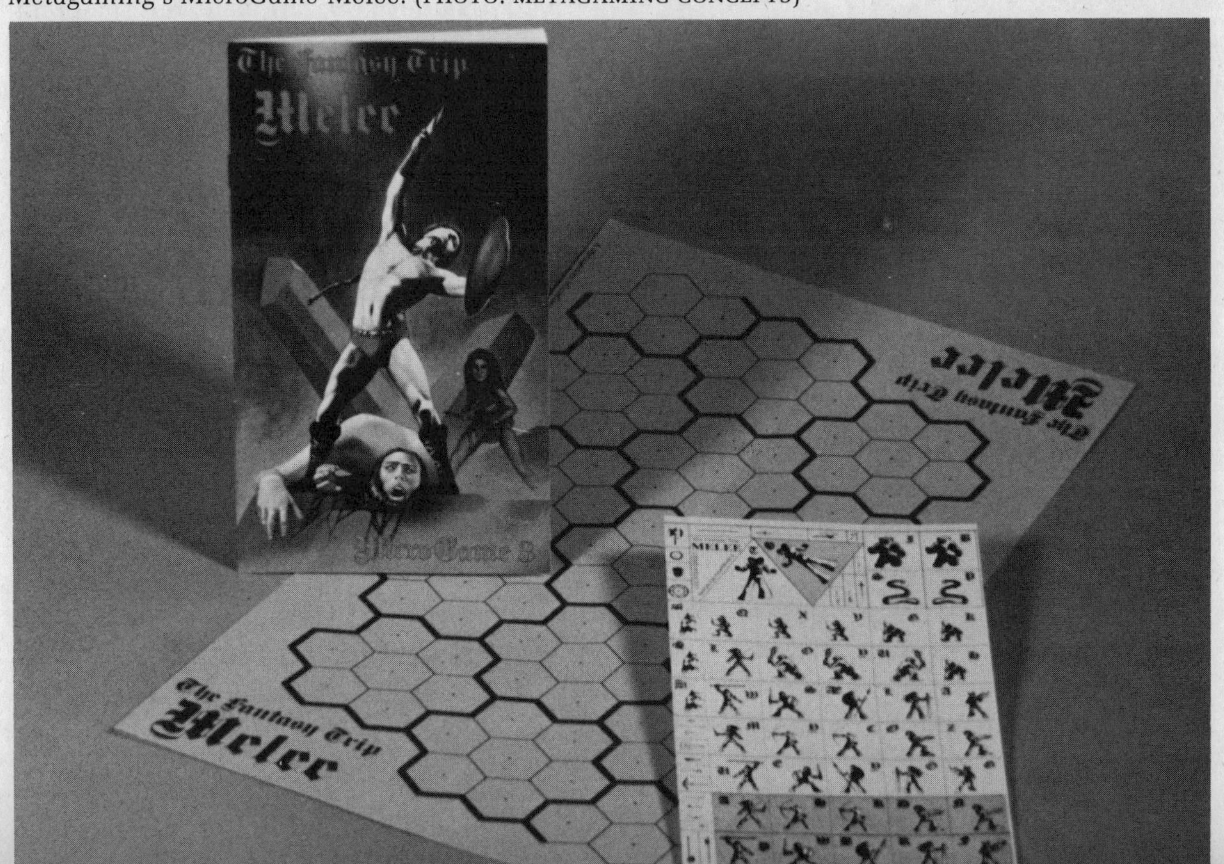

son has created the cleanest, simplest, and most understandable role-playing game around. It's also the only one that doesn't require a referee. It can be played alone, as a gladiatorial battle between men and beasts, monsters or other men, or used as an eminently satsifactory combat system within the overall framework of some other RPG. Its companion game *Wizard* supplies compatible magic rules, and either or both can be used in the solitaire dungeon *Death Test.* Both were to be incorporated into a full role-playing campaign system called *The Fantasy Trip: In the Labyrinth.* Despite its atypicality, *Melee* is probably the best place for aspiring RPG players and designers to begin.

EVALUATION:
Presentation—Good
Rules—Very Good
Playability—Excellent
Realism—Very Good
Complexity—4

OVERALL EVALUATION: Very Good

Metamorphosis Alpha (1976)

PUBLISHER: TSR Games, Inc.

SUGGESTED RETAIL PRICE: $5 (booklet)

SUBJECT: Role-playing rules are provided for a campaign set inside a giant starship.

PLAYING TIME: This is discretionary; typically, three to six hours at a sitting.

KEY FEATURES: Science-fiction fans will recognize the setting of this RPG from Robert Heinlein's *Orphans of the Sky*, Brian Aldiss' *Starship* and the unsuccessful television series, "The Starlost." An enormous colonial ship has broken down en route to another world, and the closed, terrestrial ecologies within have been subjected to massive doses of radiation. The resulting mutations among flora and fauna are responsible for a variety of monsters and the special abilities common to people and animals alike. Among the usual batch of characteristics, Constitution determines the number of hit dice a character gets. Physical and psionic combat are adapted from *Dungeons & Dragons,* although the systems aren't entirely compatible. Armor classes still run backward: high number ratings go to good weapons but poor armor. (Why does TSR perpetuate this?) There are no experience points or levels.

COMMENTS: The popularity of *Dungeons & Dragons,* a fantasy game, has made a lot of people feel they could do the same thing with a science-fiction rationale. James Ward's *Metamorphosis Alpha* is an attempt to do that by the people who brought you *Dungeons & Dragons,* and, predictably, it shares many of the parent game's flaws. The rules are better constructed, but the game takes itself too seriously. The author and publisher want you to believe the rationale, but the premise is easier to swallow than the creatures you meet: hawkoids, bearoids, cougaroids —everything but voidoids. Instead of trolls with broadswords, you get lizardmen with hand flamers. A plant like the Death Growth, which works just like Green Slime, seems to be left over from a dungeon corridor. The game is *Dungeons & Dragons* in disguise; it could as easily have been called *Missiles & Mutants.* As an honest fantasy, it would be no worse than most. As "science fiction," it comes out looking like *Gorgo Meets Star Wars.*

EVALUATION:
Presentation—Fair
Rules—Fair

Playability—Good
Realism—Poor
Complexity—6
OVERALL EVALUATION: Fair

Runequest (1978)
PUBLISHER: The Chaosium
SUGGESTED RETAIL PRICE: $8 (rule book)
SUBJECT: Rules are provided for a fantasy role-playing campaign on the imaginary world of Glorantha.
PLAYING TIME: This is at the discretion of the players and referee; three hours minimum, typically four to eight hours.
KEY FEATURES: Characters in *Runequest* begin with the usual assortment of characteristics (five of *Dungeons & Dragons'* six plus Power and Size) created in the usual way. However, there are no experience points, levels, or character classes, so characters are free to develop any skills or abilities they please, subject to the broad limits of their characteristics. A character's chances of improving a skill after an adventure are, loosely, inversely proportional to his current level of ability. He may also improve through training (by an expenditure of time and money). Spells are fairly simple and strongly oriented toward practical combat uses: helping friends and hindering foes—like *Wizard's*, only more so. Melee is quite involved, and combat requires separate rolls for attacking, parrying, damage, and hit location. Critical hits ignore the effects of armor, which otherwise reduces damage.
COMMENTS: Designer Steve Perrin has put together what might be a model for an RPG rule book: it's coherent, understandable, and full of examples to illustrate almost every point. The campaign background is a bit sketchy, perhaps, but that's a minor quibble. There are two basic difficulties to *Runequest*. The first is its cumbersome combat system, which is more realistic than most but rather tedious. If you thrive on hacking limbs in loving detail, this may not be an objection. The second it shares with *Empire of the Petal Throne*: the unfamiliarity of the world of Glorantha—the setting of *White Bear and Red Moon, Nomad Gods*, and *Runequest*—discourages emigration. Nor does *Runequest* lend itself to being grafted in pieces onto other settings; the systems are too well integrated (a novel difficulty). Classless characters, for instance, while not an unattractive concept, are not typical of most fantasy fiction or RPG campaigns and would be lost without the particular combat and magic systems. To a great extent, this means the game must be adopted entirely or not at all. Considering the legions of confirmed *Dungeons & Dragons* players, that's too much of a burden even for something this well done.
EVALUATION:
Presentation—Good
Rules—Excellent
Playability—Good
Realism—Very Good
Complexity—8
OVERALL EVALUATION: Good to Very Good

Space Patrol (1977)
PUBLISHER: Gamescience
SUGGESTED RETAIL PRICE: $6 (booklet)
SUBJECT: Game deals with personal combat in the future (within the nominal framework of an RPG).

PLAYING TIME: A game takes two to five hours.

KEY FEATURES: This is a reasonably comprehensive but understandable system for treating personal combat with science-fiction weapons and devices. With one name change, the six character attributes are straight out of *Tunnels & Trolls*, with much the same effects. There are also some interesting tables that are used to generate the shape and certain additional characteristics of humans and aliens. There is a hint of a system like *Runequest's* that gives characters a chance to increase their skills through experience.

COMMENTS: *Space Patrol* is a science-fiction version of *Melee* with pretensions. Unlike the typical TSR product, the problem is not that it takes itself too seriously; on the contrary: the authors' tongue-in-cheek approach is one of its biggest attractions. However, there is a clear implication that *Space Patrol* is supposed to be something other than just a combat system—and it isn't. Scenario generation consists of five short paragraphs of broad suggestions. The creation of entire worlds is relegated to one very brief table. There is little provision for experience and none whatsoever for spaceships, alien cultures, general background, daily existence, occupation, monetary gain, or plain ordinary motivation. As a combat system (only) incorporated into the overall framework of another RPG (like *Traveller*), this has possibilities; as the basis of a campaign, you could do as well with the game of *Monopoly*.

EVALUATION:
Presentation—Fair
Rules—Good (as a combat system)
Playability—Very Good (as a battle game)
Realism—Fair
Complexity—5

OVERALL EVALUATION: As a combat system, Good; as a game system, Poor

Superhero 2044 (1978)

PUBLISHER: Gamescience

SUGGESTED RETAIL PRICE: $7 (booklet)

SUBJECT: Role-playing rules are provided for a campaign set in a frankly implausible twenty-first century featuring the sort of superheroes who populate comic books.

PLAYING TIME: This is discretionary; typically, two to four hours at a sitting.

KEY FEATURES: This game puts you into the cape and cowl of a costumed crime fighter. Characters are divided into three types: the Unique (like the Flash) has some peculiar power or ability; the Toolmaster (like Iron Man) uses technological devices; the Ubermensch (like Tarzan) is an extraordinarily developed man-among-men. A hundred forty points are apportioned among seven characteristics (rather like *Melee* and *Wizard*), which are then modified by character type. Combat includes physical, mental, magical, and technological attacks with a multitude of modifications. Characters plot their weeks in advance (as in *En Garde!*, but in much more detail); typical activities include training and, of course, crime fighting. There is little provision for player interaction, and there are even rules for true solitaire scenarios that allow a player to be his own referee.

COMMENTS: Designer Donald Saxman obviously knows the territory. The rationale used to justify superheroes is elaborate, ingenious, appropriate, and faithful in spirit to its topic, but let's face it: Costumed comic book characters are as hard to believe as campaign promises. Here, considering the nature of the protagonists, their typical activities and opponents are disappointingly ordinary. (It may not work that way in practice, but that's how it's presented.) The combination is neither as plausible as *En Garde!* nor as exciting as *Dungeons & Dragons*. While the rules are quite detailed in many respects, they are not particularly clear or well organized. In fact, except for some marvelous illustrations, the whole thing is rather sloppily put together. It's unique, though, and comics fanatics who are already knowledgeable role-playing gamers could have fun with it.

EVALUATION:

Presentation—Fair to Good (saved by artwork)
Rules—Fair
Playability—Good
Realism—Fair
Complexity—8

OVERALL EVALUATION: Fair

Traveller (1977)

PUBLISHER: Game Designers' Workshop

SUGGESTED RETAIL PRICE: $11.98 (three booklets, boxed)

SUBJECT: Role-playing rules are provided for a science-fiction campaign set in the future.

KEY FEATURES: In *Traveller*, skills and learned abilities are more important than basic characteristics, which serve largely as limiting factors. Much time and space is devoted to the development of character background: military service, training, retirement benefits, connections. Personal combat covers everything from halberds to blasters, from military actions to starship battles. There are also rules, charts, and tables for starship construction, world generation, aliens, trading, exploration, and much else.

COMMENTS: *Traveller* is the only serious attempt to provide a really comprehensive set of role-playing rules for science fiction: interstellar travel, exploration, trade, combat at all levels, and so on. It is a conscious attempt to provide a science-fiction counterpart to what *Dungeons & Dragons* represents to fantasy gamers. Indeed, the most often voiced (if trivial) criticism of *Traveller* has been its inclusion of archaic weapons in its combat tables. Considering the territory it seeks to cover, *Traveller* is necessarily complex, and it presumes on the part of the potential referee considerable familiarity with other role-playing games and the literature of science fiction. While neatly presented, it ignores playing hints and examples of play. For experienced players wishing a truly open-ended, science-fiction, role-playing campaign, however, there is no real alternative.

EVALUATION:

Presentation—Good
Rules—Good
Playability—Fair
Realism—Very Good
Complexity—8

OVERALL EVALUATION: Good

Tunnels & Trolls (1975)

PUBLISHER: Flying Buffalo, Inc.

SUGGESTED RETAIL PRICE: $5 (for one booklet)

SUBJECT: Fantasy role-playing campaign rules are provided.

PLAYING TIME: This is discretionary; typically, two to six hours at a sitting.

KEY FEATURES: *Tunnels & Trolls* is a less complex *Dungeons & Dragons.* Characters are generated in the same way, but Wisdom has been replaced by Luck, and there is no separate priest class. Experience points and levels are similar to *Dungeons & Dragons* (with equally silly names for the individual levels), except that increases in level allow progressively larger increases in one or more attributes. Sustainable damage (what is called "hit points" in *Dungeons & Dragons* but *not* in *Tunnels & Trolls*) is based on Constitution. Combat is more abstract: in brief, dice are rolled for the weapons involved (bigger weapons—more dice), with additional points given for individuals with exceptional Strength, Dexterity, or Luck. The group with the lesser total of *Tunnels & Trolls* style "hit points" takes the difference in the totals as damage distributed among the individuals in the group. Armor absorbs some damage. Spells are dependent on IQ and Dexterity and, temporarily, cost Strength to cast because they fatigue the user. Only six-sided dice are required.

COMMENTS: *Tunnels & Trolls* was specifically designed to be a cheaper and simpler *Dungeons & Dragons.* It's both—and faster—but not better. Several new editions over the last few years have added a few suggestions and clarifications, but there are still nearly as many holes in the rules as there were in *Dungeons & Dragons.* There are no monster or treasure lists, no thiefly abilities; everything, including the quick-and-easy combat system, seems designed to let you "get right down to the game"—but what's left? Some good ideas for the treatment of weapons and armor are never developed. Combat is unsatisfyingly gross for a personal- (man-to-man) level game and cannot treat many details at all adequately without being bent out of shape. The spells have unbelievably tacky names—Too-Bad Toxin, Yassa-Massa, Dum-Dum, Upsidaisy—which lend the game an air of *Ali Baba and the Three Stooges.* Increasing character attributes *wholesale* results in "human" characters with an IQ of 450 or the strength of King Kong. That's not fantasy; it's nonsense. There are better alternatives available.

EVALUATION:
Presentation—Fair
Rules—Poor
Playability—Good
Realism—Poor
Complexity—6

OVERALL EVALUATION: Fair

Wizard (1978)

PUBLISHER: Metagaming

SUGGESTED RETAIL PRICE: $3.95 (plastic envelope)

SUBJECT: Role-playing rules are provided for magical combat.

PLAYING TIME: It takes thirty minutes to one hour.

SCALE: This is a close-tactical game; each counter is an individual man, animal, or monster. Hexes represent an area about six feet across, and each turn represents a few seconds.

SIZE: The 12″ x 12″ "board" uses 25mm (1″) hexes.

BALANCE: This is excellent.

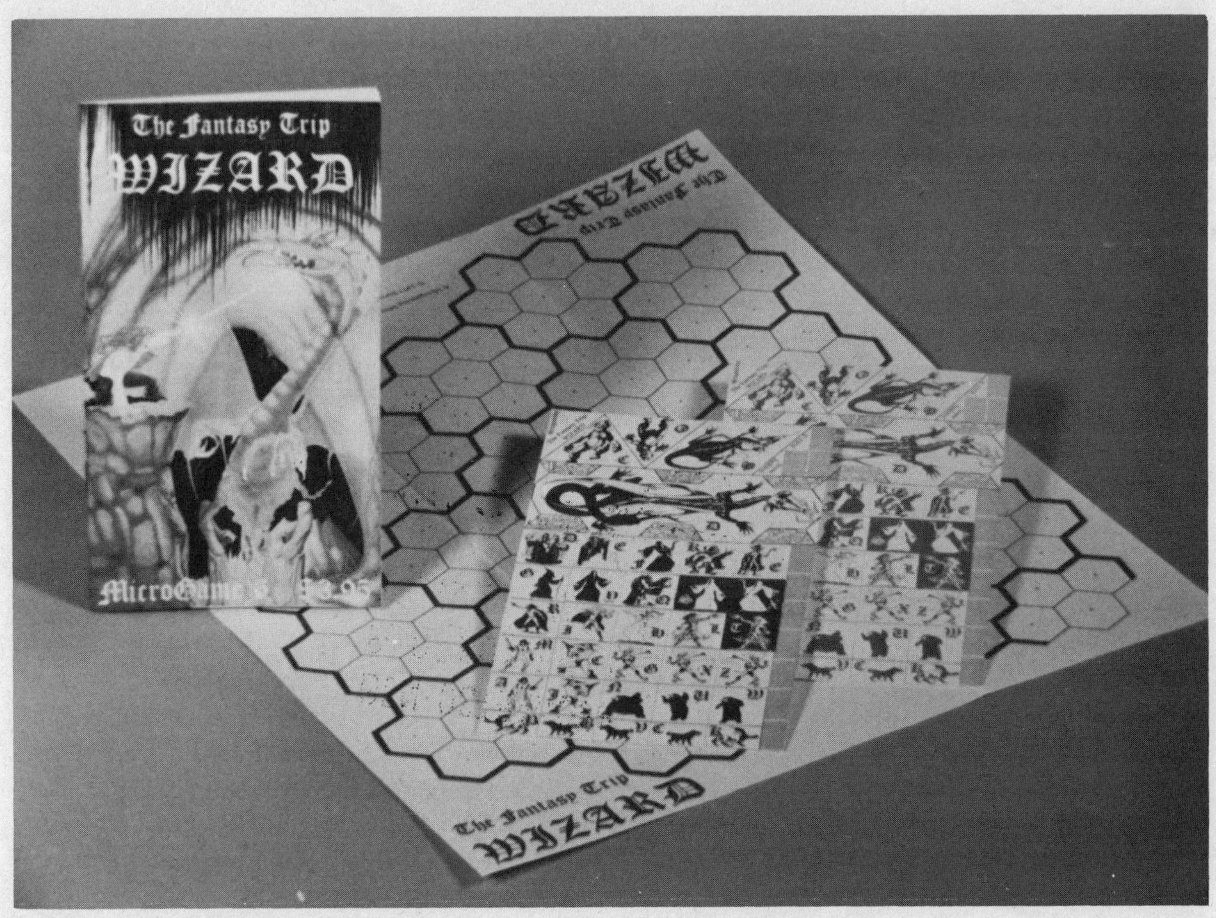

Metagaming's MicroGame Wizard. (PHOTO: METAGAMING CONCEPTS)

KEY FEATURES: This is the magic system for *Melee's* combat system. To the other game's Strength and Dexterity is added a third attribute, Intelligence (IQ), which equals the number of spells that a wizard knows. It also controls how complicated a spell he can learn. There are four classes of spells—Missile Spells, Thrown Spells, Creation Spells, and Special Spells. Rules govern each class. There is not quite the variety found in some systems, but spell use and effects are clearly and reasonably delineated. As in *Melee*, there are no "levels," but players can trade in experience points for increases in attributes. (Unlike *Tunnels & Trolls*, however, these "purchases" are retail rather than wholesale.)

COMMENTS: *Wizard* is really the completion of *Melee*. Though it is playable in its own right (as is *Melee*), the whole is greater than the sum of the two parts. A fantasy game is nothing without magic, and a magic game lacks flavor without some hulking barbarians. The system implicit in these two games is commendable. It's a basic RPG framework without half of the pseudophilosophical chintz in others. With the microgame reputation for simplicity and economy, it may be too good a bargain to be taken as seriously as it deserves. If so, that oversight should be remedied when *The Fantasy Trip: In the Labyrinth* is available.

EVALUATION:

Presentation—Good

Rules—Excellent

Playability—Very Good

Realism—Good

Complexity—5

OVERALL EVALUATION: Very Good

CHAPTER 17

Computers and the Future of Wargaming

MORE GAMES are played on computers than you might think. People are commonly introduced to the use of home computers by means of games, and novice programmers inevitably direct their first creative efforts to developing a new version of a game like *Star Trek* or blackjack. In fact, professional programmers, systems analysts, and computer operators spend so much of their time playing sophisticated games like *Zork* or *Adventure* on corporate machines that elaborate codes, locks, time limits, and identification procedures have been instituted by many companies to discourage this practice.

The computer-run simulations of Pentagon and civilian think tanks are the technological descendants of the military wargames of the 1824-to-1945 period. After being fed masses of data programmers consider relevant, a highly sophisticated computer then plays a "game"—or, more frequently, a series of games—in which it takes the part of the referee and two or more sides. Because such computers are capable of performing very involved calculations rapidly, it's possible to test the effects of a great number of variables—anything from a higher-than-anticipated desertion rate in a particular regiment in West Germany to the effect on worldwide food production of a two percent average drop in temperature. Since the subjects of these simulations are hypothetical, they are less re-creations of history than what might be termed "pre-creations."

Because such simulations frequently involve subjects many people would rather not think about, like the results of a nuclear conflict, some contend that serious conclusions about such vital matters should not be drawn from a "game"—a thing-which-is-not-real. However, it's not clear that today's planners regard their military wargames any more seriously than yesterday's German generals and Japanese admirals. Moreover, despite their limitations, games or simulations are better than nothing, particularly when few people find the alternative—field testing—attractive! Whatever their uses, they have little direct bearing on commercial wargaming.

The first computer games widely known

to the public were arcade video display devices of two types. The first were action games, like electronic table tennis played in "real time": that is, there was a one-to-one correspondence between the timing of events on the video display screen and those at the controls. The electronic paddle moved as fast as you turned the knob *when* you turned the knob. The same principle applied to the maneuvering and weapon "firing" of a tank, biplane, or spaceship. In addition to being limited to a single weapon or paddle per player, such games required quick thinking of a limited sort but rewarded reflexes and coordination more than cerebration.

Devices of the second type were, in effect, elaborations of tick-tack-toe. These were always solitaire electronic games. Some were contests, battles of "wits" between computer and player. These were standard abstract games with simple rules or newly devised games based on numbers or geometry—subjects that the computer could handle easily. In other cases like blackjack, the computer acted both as dealer and card deck; the player might be engaged in a contest with a simulated opponent but more commonly his opposition was simply the randomness of chance, as in any other solitaire game. Except for showing that a computer could "play the part," many of these games had little reason to exist as *computer* games. Even the others were far closer to Chinese checkers or, at best, go than to *Tactics II*.

RHYMES AND REASONS

These games, however, represent only a fraction of the possibilities offered by the computer. Action and movement—rockets blasting, lasers flashing, explosions thundering—have their attractions; they allow an arcade space wargame to be more vivid, more lifelike, in its limited way, than the most elaborate conflict-simulation board game. To the degree that such effects can be integrated with more complicated, mentally challenging games, they will be far more pleasing than consulting a Combat Results Table and flipping over a counter.

With the use of computers, rule disputes, page-turning, and chart-checking may become things of the past. Although there may still be a need for a rule book of some sort—either a printed version or a video display on the screen—many of its functions will be taken over by the computer program itself. If a program is properly "debugged," the computer will be able to enforce the rules impartially, preventing cheating, playing out of turn or sequence, differences of interpretation, and simple mistakes. More significantly, perhaps, many of the "rules" need not be learned by players at all; they can be built in. For instance, in *Starfleet Orion*, a wargame designed for use with microcomputers, a starship may fire at an enemy craft simply by designating a power level for its weapon beam. But because of computer decree, the beam cannot exceed the power available, the designed strength of the beam, or its current state of repair. Taking into account the accuracy of the ship that is firing and the distance and size of the target, the computer first determines if a hit was scored. If so, it then modifies the strength of the striking beam according to the distance between the two ships (and, of course, its initial power), subtracts whatever protection might be afforded by the target ship's armor and energy shield level, and allocates the net damage among the target ship's various subsystems. In a conventional wargame, such a sequence would require pages of rules, charts, die modifications, and a good deal of die-rolling. With a computer, they aren't needed, and the task is performed in a fraction of the time it would take to do "manually."

Complicated calculations and bookkeeping required in many wargames are a tedious annoyance and the chief limiting factors in both role-playing games and the multiplayer games of interstellar exploration and conquest. What's my production level this turn? When do my tanks come off the assembly line? How many hits on the *Mary Deare*? Where are my "hidden" units? And so on. A computer can keep track of all those things and display them on its video display screen at the touch of a key. Aside from eliminating errors of calculation and reducing errors of entry, a computer can also introduce subtle modifications that were too complicated to figure "by hand" and can allow many more options.

Computers can moderate multiplayer games in much the same way nineteenth-century referees handled *kriegspiel*, games masters coordinate play-by-mail *Diplomacy* games, and dungeon masters run contemporary role-playing games. One of the biggest advantages of the computer moderator —as with a human referee—is that it allows the thrill, mystery, and challenge of limited intelligence: a player may be kept somewhat in the dark as to the location, identity, moves, and intentions of his opponents. While this is appropriate to many—perhaps most—wargames, it's almost essential to simulations of science fiction and fantasy, in which such factors would certainly be unknown and possibly unsuspected.

The computer is also potentially the opponent that is always available. Many games are played solitaire not because most gamers prefer to play alone, but because they cannot find suitable opponents: those who like to play the same games, those who are at the same approximate skill level, and those who have the time available when the gamer has the time and inclination to play. Time sharing on large computers doesn't solve all these difficulties, but a home or personal microcomputer is ready for a game at the flip of a switch. It never gets drunk, restless, or angry at you; nor does it blow smoke in your face. It can be beaten fifty times in a row without bearing a grudge or dampening its enthusiasm. It can even be programmed to play at different levels to accommodate growing skill—or the vindictiveness of your mood.

Finally, computers are an open invitation for you to design your own games. As a design-a-game kit, even a microcomputer makes the average role-playing rule book look limited and limiting. This urge to participate is clearly responsible for the proliferation of quite mediocre computer games currently on the market, but in the long run it will lead to higher standards than even the board wargame field is used to. And it will always provide a unique source of enjoyment for anyone who wants to turn loose his powers of creativity.

WELCOME TO MY MORNING

Integrated circuits (ICs) that are the size of the average wargame counter now do the work of the console-sized computers of a decade ago. In addition to making possible powerful, typewriter-sized microcomputers, they have created a boom in inexpensive "computerized" games like *Simon, Merlin*, and hand-held pinball games. And they are adding a new dimension to board games like *Stop Thief* and *Code Name: Sector.* For example, *Code Name: Sector* is a simplified wargame involving four destroyers (moved by up to four players) in pursuit of a computer-controlled hidden submarine. The computer keeps track of the speed and position of all five vessels, their range from each other, and the effects of collisions, depth charges, and missed shots. The submarine's pattern of movement is simple, and its evasive capacities are limited, but five years ago the game couldn't have existed at all. As the cost of electronic circuitry continues to drop and competition increases, such games will become cheaper and more sophisticated, but their target audience—children and young adults who know nothing of simulation games—will probably keep them from having more than incidental interest to real wargamers.

Real-time action games are certainly as popular at arcades as ever, and every year sees more complicated and more engrossing versions of Tank and Space War. There are now variably selective space wargames that realistically handle vector mechanics, inertia, acceleration, and the effects of gravity. The capabilities of these machines' displays, controls, and data processing are obviously far greater than something selling on the shelf of a discount store for $24.95.

At-home stepchildren of arcade games, the domain of Atari and competing companies, have grown increasingly sophisticated and expensive. As the price of "real" computers has fallen, the line separating computer game systems from microcomputers with game software grows thinner almost by the month. In a year or two, at most, it may disappear; however they are marketed, "game-only" computer systems will cease to exist as something separate from true programmable microcomputers. However, they will have done their part to lure hundreds of thousands of people to consider small computers as game devices.

LEAVING, ON A JET PLANE

For at least the next five to ten years, the technological future of wargaming lies with the microcomputer. It has been estimated that about twenty-five percent of the wargamers in the United States have one. Compared to the separately packaged computerized games like *Stop Thief*, the microcomputer offers enormous advantages in flexibility, power, and—in the long run—price. You only have to buy the computer once, and upgrading its capabilities with more memory, a disk drive or a printer is comparatively inexpensive (if not cheap). Compared to time sharing on a large computer, the incremental or marginal cost of playing games on a microcomputer is minimal: you have already bought the whole package, and, psychologically, it's over and done with.

There are at least five distinct types of games or wargaming uses that can be accommodated by microcomputers. The simplest is the one- or two-player real-time action game. Current versions are fun to play, but they aren't real wargames—yet. There is perhaps no way around the basic limitation of a single tank, ship, or plane per player, but within that scope the degree of realism and the options available—the amount of control exerted—could certainly make such games an attractive alternative to, say, *Sniper!* or the simpler scenarios of various naval games.

A step up or over is a genuine two-player tactical space wargame like Automated Simulations' *Starfleet Orion*. It's not a real-time game; it's a contest of strategy and tactics like any board wargame. Each of two players controls up to nine ships on a side, each of which has independent movement and combat capabilities, including energy beams, tractor/pressor beams, defensive force shields, missiles, and torpedoes. Twelve scenarios are included along with provisions for designing both new ships and new situations. The computer handles the calculating, bookkeeping, and "die rolling" and permits the use of limited intelligence and simultaneous movement and combat. Although hexes are not used, there is a positional grid to regulate movement and ranged combat. While this kind of thing has wide application to wargames of any setting or period, it's not effective with large numbers of units; a standard display screen isn't large enough or the resolution fine enough to yield the information required without using some sort of split display, which becomes hopelessly tedious with the equivalent of hundreds of counters.

Solitaire games in which the computer furnishes the opposition are most obvious and perhaps more desirable than any other sort. Ideally, solitaire games should be as sophisticated and challenging as an ordinary wargame. Early ones were more limited. Cybernautics' *Trek '78*, a microcomputer version of what is, in all its variations, almost certainly the most widely played computer game in history, is a pleasant but not terribly elaborate simulation of the Good Ship *Enterprise*'s never-ending battle for truth, justice, and NBC—er, the Federation. Similar struggles between the graceless battlestar *Galactica* and the robot Cylons are probably inevitable—unfortunately. Armed with the familiar battery of phasers, shields, photon torpedoes, and a large but limited (and nonrenewing) amount of energy, the U.S.S. *Enterprise* travels around a square segment of the galaxy in an attempt to eliminate as many Klingons as possible before going back to the starbase to refuel. The player's choices are few, and the computer opposition is quite unsophisticated: the Klingons will fire back if not destroyed, but won't move on their own accord.

A "second-generation" computer opponent can be found in Automated Simulations' *Invasion Orion*, in effect a sequel to *Starfleet Orion* with ten new scenarios designed for solitaire play. In addition to the same referee functions that it performed in *Starfleet Orion*, *Invasion Orion's* program enables it to run the ships of a hostile race of robots called, appropriately enough, the Klaatu. Except for the lack of tractor beams, the computer side's options and capabilities are exactly the same as a human player's: movement and four weapons systems for up to nine ships (and/or planets, asteroids, space stations, or whatever) *against* up to nine craft. The computer can be set to one of three levels of playing ability and is a very worthy opponent indeed. Even more sophisticated, artificial-intelligence opposition is possible with memory capacities of greater than 16K (16,000) bytes of memory, the standard size of Radio Shack's TRS-80 Level II microcomputer, which is one of the most popular in the industry.

Closely related are computerized, one-player role-playing games. *Zork* and *Adventure* are extremely elaborate examples, but they exist only on very large computer systems. In both, a single adventurer wanders through a preconstructed (and fixed) "dungeon," overcoming various obstacles by fighting, thinking, or making use of often seemingly trivial items that are found, stolen, or taken from separate locations. These games are marvelous fun, mentally challenging, and far superior to noncomputer "flow-chart" solitaire dungeons.

While these games don't quite permit the range of activities allowed in standard role-playing games, their options and vocabulary are extraordinary. One of the particular joys of playing *Zork* or *Adventure* is that the flexibility of their language allows actions and replies to be stated in quite normal English. For example:

I'M GOING TO OPEN THE WINDOW.
IT WON'T OPEN.
THEN I'LL BREAK IT.
WITH WHAT?
MY FIST.
THE WINDOW BREAKS BUT YOU CUT YOUR HAND.

The use of "normal" language makes the games easy to play mechanically but very difficult intellectually. There are no rules—no rigidly limited set of commands—to suggest what can or can't be done with or to any particular object. The range of possible or appropriate behavior is not by any means infinite, but it is far larger than can be comprehended in hours of playing.

Unfortunately, nothing so elaborate or open-ended is possible on microcomputers with memories of a few thousand bytes. Automated Simulations' *DunjonQuest* series, by far the most sophisticated of the microcomputer role-playing games, allows only a handful of basic commands. On the other hand, each game has four levels of about sixty rooms, chambers, or corridors apiece, dozens of monsters and traps, and plenty of treasure and magical items. Added to the excitement of real-time action and further subprograms that allow meaningful—and lively—negotiations for purchasable goods, these games, too, compare favorably with printed solitaire dungeons and allow varied and challenging solo adventures.

Still larger and more complicated dungeons, more commands and options, and a system that allows the general and varied use of magic all depend on greater computer memories, but that's just a matter of time.

A final use of microcomputers is not really as a game at all but rather as a game aid. There are programs that perform many or all of a dungeon master's calculations and bookkeeping for him. By keeping track of hit points, endurance, fatigue, experience points, damage, "die-rolling," and the like, such a program can speed up role-playing games immensely. By reducing time and tedious tasks, two of the biggest drawbacks of running an RPG, the computer makes it more enjoyable for referee and players alike. Still being developed are even more elaborate versions that would require at least 32K to 48K bytes of memory, but that could conceivably "create" wandering monsters and even random dungeons.

The same functions would be an aid to many standard wargames, and indeed Simulations Publications has concentrated its initial microcomputer efforts on game aids that might incorporate Combat Results Tables, production cycles, die-rolling, and the like. While this sort of thing could surely be useful, one must question how much of a market there will be for a piece of microcomputer software sold at any realistic price that will only be relevant to one wargame. Unlike role-playing campaigns, which may be played every week, the average wargame may not be played half a dozen times. One must also question why SPI is beginning with aid programs for some of its least successful games, such as *Agincourt*.

FAREWELL ANDROMEDA

Although microcomputers are hogging the spotlight, they are not the only act under the gaming tent. The major commercial use of large computers in games is to moderate multiplayer games. Although other firms threaten to get into the act, the only company of note operating such games at this writing was Flying Buffalo, Inc., whose major game in this area is the science-fiction game *Star Web*.

The mechanics of such an operation are similar to those of a normally moderated play-by-mail game of *Diplomacy*. Every two weeks, or at some such interval, each player receives a computer printout of his status in the game: where his ships are; how his production levels stand; how many points he has; and what he knows about the planets, ships, and other players in the game. Within a few days he sends in his orders, which must conform to the coding requirements of the program—artificial but not difficult to understand or use. The economics of it are even simpler: each player pays a fixed amount for every turn he's in the game.

Star Web is clearly a cousin of *Outreach* and *Stellar Conquest*, but it has the advantages a computer moderator affords. The appearance of the map; the number of planets, their resources, and the connections between them; the number, location, and intent of other players: all these are unknown to the individual player in the beginning. There is a common goal—a certain number of points—but the scoring varies according to the race or role taken by the player (and which is also hidden from his opponents). For example, a Merchant gets points for trading; a Pirate gets points for plundering worlds; an Apostle gets points for converting the populations of planets to his point of view; and so on. Wars, alliances, or trading between players are all possible. If you like games of the exploration-and-conquest variety, this is the way to go.

While the subjects of most existing computer games are science fiction or fantasy, this need not be the case. SPI is planning a massive computer-moderated simulation of the Hundred Years' War, although some observers feel that a nearly unknown conflict in a period manifestly unpopular with wargamers isn't the best of subjects for such a project. Of course, SPI has fooled the oddsmakers before, and it may again.

POEMS, PRAYERS, AND PROMISES

At the moment, Automated Simulations has a near monopoly on sophisticated role playing and wargames for microcomputers. This can't last. Avalon Hill has contracted with a midwestern firm to supply it with computer games, and Metagaming has long shown an interest in translating some of its games into a computer format. Most of the other major wargame companies have taken at least preliminary steps toward joining the boom in electronics. That they haven't moved faster and accomplished more is partly a product of their size and partly of a certain evident confusion as to the proper uses of computers for gaming; for most of them, it's a whole new field. They will get untracked eventually.

In five years it is not unlikely that a major portion—perhaps the distinct majority—of all wargames will be computer-based or -aided. The computers in question may be ever more powerful micros or they may be tiny minimicros—fancy programmable calculators—used as play aids, but they'll be there in some form. The advantages are too great, and the challenges too compelling, to make it otherwise.

In the more distant future, if the universal telephone/television/stereo/computer network becomes a reality, at-home multiplayer games, in real time or not, will probably become commonplace. If such network systems replace individual independent microcomputers, there is almost no limit to the sophistication of possible games, and computer opponents could be the equal of human players in any of them.

Technology may impinge on wargaming in quite different ways. The U.S. Army has used low-powered lasers and "computerized" reflector vests to play life-size wargames, and a Los Angeles firm is trying to turn this concept into a gaming arena for the ultimate *Sniper!* Though some people have objected to this on moral grounds, the idea of "live" games is no different from the Society for Creative Anachronism's recreations of medieval combat and revels and no different in principle from the cowboys-and-Indians (or, these days, Jedi-knights-and-storm-troopers) games that most children play.

Eventually, life-size simulations and computers are going to combine to make the ultimate game form: the total-participation environment somewhat foolishly exemplified by the movie *Westworld*. Although stupidly dangerous mechanisms and impossible events were dictated by a contrived plot in the film, the idea of a somewhat more sophisticated Disneyland is not so farfetched and is not unknown elsewhere in science fiction. Although freestanding, mobile, truly manlike androids are considerably beyond present-day technology, and sexually attractive ones (despite *Westworld* and *The Stepford Wives*) are not just silly but grotesque, mechanical dinosaurs of limited action are only a small step away from what the Jungle Cruise through Frontierland already affords. Arcade fighter-pilot and space-war games are only a small step from NASA's lunar lander simulator. A computer-generated scenario could divide human participants into sides or teams with differing objectives, and period costumes and safe and sophisticated "weapons" could allow anyone the chance to act out his wildest fantasies—to *really* play the part of Captain Blood or Luke Skywalker or, for that matter, Princess Leia.

In fact, the biggest requirements for such an amusement park are not technological at all. What is needed, primarily, is a recognized market for such elaborate and expensive games and a social climate that accepts or even encourages the acting out—the playing—of such fantasies. The success of Disneyland, Disney World, and the television series *Fantasy Island* indicates that those factors may be already present.

There is every reason to believe that games of all sorts—simulation games, wargames, "new" games, total-participation fantasies—more challenging, more sophisticated, more elaborate, and more lifelike than ever—will be a big part of our future.

Appendix

Glossary

Advance after combat—A move allowed to the "victorious" unit after a defending unit is eliminated or forced to retreat; it takes place at the end of the combat phase of a turn (not the movement phase). It does not count as part of the unit's normal, per-turn movement limitations.

Attack factor—A number that signifies the comparative strength of a unit when attacking opposing units.

Breakdown—Substituting two or more smaller units (e.g., battalions) for one larger unit (e.g., a regiment). (Compare *integrity*.)

Combat factor—A number that signifies the comparative strength of a unit when engaged in combat with opposing units (either on attack or defense); a combined attack and defense factor.

Combat Results Table (CRT)—A chart or table that is used to resolve combat, usually with the aid of a die. There are two main types: An *odds/ratio* CRT is divided into columns according to the ratio of the attacker's combat factors to the defender's combat factors. A *differential* CRT is divided into columns according to the numerical difference between the opposing units' combat factors. A less common third type, found only in tactical games, is based solely on attack factors without regard to defense.

Command control—The ability (or inability) to make units do what their leaders (i.e., you, the player) wish. Command control rules usually involve ways to prevent certain (often randomly selected) units from "following orders."

CRT—Combat Results Table.

Defense factor—A number that signifies the comparative strength of a unit when being attacked by opposing units.

Double-impulse movement—Two movement phases, one before and one after combat; the second phase (known sometimes as *mechanized movement*) is usually restricted to certain motorized or armored units.

Dummy counters (or units)—Fake or decoy units that are only used to deceive an opponent as to the actual dispositions of a player's forces; generally used in conjunction with *inverted counters*.

Facing—In tactical games, the differentiation of a unit's front and rear ends (also, sometimes, its sides or "flanks"); also, therefore, the direction a unit is pointing (which will usually affect where it can move or fire).

Factor—A number that signifies the comparative ability of a unit to perform certain tasks, especially of combat or movement; also, loosely, a single point or increment of such a rating.

Geomorphic—See *isomorphic*.

Hex—One hexagon in a hexagonal grid.

Hex-side—One of the six sides of a hexagon.

Hidden movement—Movement of units in such a manner that the opponent is unaware of their exact location and/or composition; this may involve two boards, *movement plots, inverted counters*, or other devices.

Indirect fire—In tactical games, a type of *ranged fire* that is not subject to *line-of-sight* restrictions but that may require *spotting* or "visual observation" of the target by other friendly units.

Integrity—The ability of units of a certain size (e.g., brigades) to gain a bonus in movement or combat when combined or stacked into the next larger size (e.g., divisions).

Inverted counters—Counters initially placed or moved upside-down to conceal their identity from another player.

Isomorphic—Modular game boards that can be placed next to each other in varying orders and attitudes to alter the composition of the playing surface from scenario to scenario or (in air, sea, and space games) to allow an effectively unlimited playing area; generally found only in tactical games. Also known as *geomorphic* boards.

Line of sight—In tactical games, a line between two units or locations that determines whether one unit can "see" (and, hence, attack) a distant opposing unit. Various *terrain* features such as woods or mountains may block a line of sight.

Mechanized movement—A bonus given certain units (e.g., armor), most commonly a second-movement phase after combat for those units only; the second half of *double-impulse movement.*

Morale—The ability of units to remain in combat and/or "follow orders" until they are eliminated (as opposed to "fleeing" or becoming "demoralized"); also, a rating or factor connected with such an ability. (This is not the same as *command control.*)

Movement factor—A number that signifies the distance a unit can move; loosely, the number of hexes a unit can move in a turn.

Movement plot—A written set of orders that includes the exact hex-to-hex path a unit will follow for the current turn; also, an abridged form of this. Used with *SiMov* systems.

Operational—A game scale that is between tactical and strategic; a game whose units are battalions, regiments, brigades, and/or divisions.

Order of battle—A list of the forces by organization and type that are actually present at a battle; the units in a game or scenario.

Overrun—An overwhelming attack that is made on a unit in the same hex (not an adjacent one), often in the movement phase; this is sometimes allowed only if the attack is so strong that it cannot fail.

Plot—A written set of orders concerning the movement or attacks of a unit for the current turn; used with *SiMov.*

Power politics—Refers to multiplayer games whose concerns are as much political—negotiations, alliances, and so on—and/or economic as strictly military: for example, *diplomacy.*

Ranged fire—In tactical games, an attack that is made from a distance by archers, artillery, infantry rifles, and so on. A nonadjacent attack.

RPG—Role-playing game.

Scenario—One of a set of situations complete with its own unit listings (order of battle), victory conditions, and (sometimes) special rules that is nonetheless a part of a larger game; in effect, a minigame sharing a board, pieces, and rules with other minigames.

Sequence of play—The exact order in which things take place during a game turn.

SiMov—Simultaneous movement; one of various systems usually involving written *plots* in which movement and/or combat results take effect simultaneously, as opposed to the normal sequence in which players alternately move and fight.

Spotting—The act of a unit "seeing" an opposing unit, often so that a third unit may fire at the opposing unit. (See *line of sight, ranged fire,* and *indirect fire.*)

Stacking—Placing more than one unit (usually on the same side) in a hex; there are usually limitations on this, and in many games it is prohibited.

Strategic—A game scale that is characterized by large distances and broad concerns; a game whose units are divisions or armies (or the equivalent). The extreme case is called "grand strategic."

Supply—Considerations of ammunition, food, fuel, and so on, usually treated abstractly; unsupplied units are usually restricted in their activities.

Tactical—A game scale that is characterized by

small units (battalions or smaller) and usually *ranged fire;* the two extremes are called "close tactical" (involving individual men or vehicles) and "grand tactical" (large battles fought at this scale or those with relatively large units).

Terrain—The natural or manmade features of a hex, as indicated by its color or coding: for example, rivers, woods, seas, mountains; also, the effects of such features on movement or combat.

Terrain effects chart—A chart that explains the effects of terrain on movement and combat of units.

Untried units—Units whose actual combat strength is unknown to either player until they take part in their first battle, at which point their true combat factors (on the underside of the counter) are revealed.

Zone of control (ZOC)—A unit's "sphere of influence," usually the hex it is in and the six adjacent hexes that affect opposing units. Effects vary greatly but usually involve combat, movement, and/or supply.

Wargame Index

Wargame Publishers' Directory

Automated Simulations, P. O. Box 4232, Mountain View, CA 94040
Avalon Hill Game Co., 4517 Harford Rd., Baltimore, MD 21214
Battleline Publications (See Heritage Models)
The Chaosium, P. O. Box 6302, Albany, CA 94706
Conflict Game Co. (See Game Designers' Workshop)
Discovery Games, P. O. Box 3395, St. Paul, MN 55165
Excalibre Games, Inc., P. O. Box 29171, Brooklyn Center, MN 55429
Fantasy Games Unlimited, Box 182, Roslyn, NY 11576
Flying Buffalo, Inc., P. O. Box 1467, Scottsdale, AZ 85252
Fusilier Games, 27 Ashvale Place, Aberdeen, Scotland
Game Designers' Workshop (GDW), 203 North St., Normal, IL 61761
Gamescience, 01956 Pass Road, Gulfport, MS 39501
Gamma Two Games, Ltd., P. O. Box 46347, Vancouver, B.C., Canada V6R 4G6
Heritage Models, Inc., 9840 Monroe Dr., Bldg. 106, Dallas, TX 75220
Historical Alternatives, 1142 South 96th, Zeeland, MI 49464
Historical Perspectives, P. O. Box 343, Flushing Station, NY 11367
House of Games, Inc., 2633 Greenleaf Ave., Elk Grove Village, IL 60007
Imperium Publishing Co., P. O. Box 9854, Minneapolis, MN 55440
Marshal Games, 8604 Via Mallorca Dr., La Jolla, CA 92037
Metagaming, P. O. Box 15346, Austin, TX 78761
Operational Studies Group, 1261 Broadway, New York, NY 10001
Simulations Canada, P. O. Box 221, Elmsdale, Nova Scotia, Canada BON 1MO
Simulations Publications, Inc. (SPI), 257 Park Avenue South, New York, NY 10010
TSR Games, Inc., P. O. Box 756, Lake Geneva, WI 53147